SpringerBriefs in GIS

SpringerBriefs in GIS present compact, concise summaries of cutting-edge research, practical applications and visualizations in the use of geographical information systems. At 50 to 125 pages (approx. 20,000 – 50,000) words, SpringerBriefs in GIS provides researchers and practitioners with an innovative venue to present work that might be longer and more complex in scope than a journal article.

This series aims to cover a wide range of topics related to geographical information science and geographical information systems. Potential topics could include: an in-depth case study on the use of GIS to accomplish a specific goal; a guide to an emerging GIS tool, technique or map; a "hot-take" on or snapshot of a current issue that needs to be published as quickly as possible (just 8-12 weeks after a manuscript's delivery and acceptance). Multidisciplinary studies are particularly welcome.

SpringerBriefs are distributed through the same channels as Springer's book content, and are available as physical books and full and chapter-wise eBooks. Both solicited and unsolicited manuscripts are considered for publication in this series. Please send questions and proposals to Zachary Romano, Associate Editor, Earth Science, Environment, and Geography, at Zachary.Romano@Springer.com.

Priyanka Biswas • Nilanjana Das Chatterjee

Crime Prediction Using GIS and Statistical Modelling

A Study on Crime Against Women in West Bengal

Priyanka Biswas
Department of Geography
Vidyasagar University
Midnapore, West Bengal, India

Nilanjana Das Chatterjee
Department of Geography
Vidyasagar University
Midnapore, West Bengal, India

ISSN 2367-010X   ISSN 2367-0118   (electronic)
SpringerBriefs in GIS
ISBN 978-3-031-81447-1   ISBN 978-3-031-81448-8   (eBook)
https://doi.org/10.1007/978-3-031-81448-8

© The Editor(s) (if applicable) and The Author(s), under exclusive license to Springer Nature Switzerland AG 2025

This work is subject to copyright. All rights are solely and exclusively licensed by the Publisher, whether the whole or part of the material is concerned, specifically the rights of translation, reprinting, reuse of illustrations, recitation, broadcasting, reproduction on microfilms or in any other physical way, and transmission or information storage and retrieval, electronic adaptation, computer software, or by similar or dissimilar methodology now known or hereafter developed.
The use of general descriptive names, registered names, trademarks, service marks, etc. in this publication does not imply, even in the absence of a specific statement, that such names are exempt from the relevant protective laws and regulations and therefore free for general use.
The publisher, the authors and the editors are safe to assume that the advice and information in this book are believed to be true and accurate at the date of publication. Neither the publisher nor the authors or the editors give a warranty, expressed or implied, with respect to the material contained herein or for any errors or omissions that may have been made. The publisher remains neutral with regard to jurisdictional claims in published maps and institutional affiliations.

This Springer imprint is published by the registered company Springer Nature Switzerland AG
The registered company address is: Gewerbestrasse 11, 6330 Cham, Switzerland

If disposing of this product, please recycle the paper.

Foreword

In the rapidly evolving landscape of crime prevention and justice, the urgency of applying advanced analytical methods and technologies can hardly be overstated. The book *Crime Prediction Using GIS and Statistical Modelling: A Study on Crimes Against Women in West Bengal* stands at the intersection of these transformative approaches, offering a comprehensive exploration of how Geographic Information Systems (GIS) and statistical modelling can be harnessed to address one of the most pernicious and pressing issues of our time: crimes against women.

Despite numerous measures and laws in place, crimes against women are pervasive, cutting across cultures and geographies. Occurrences of gender-based crimes continue to skyrocket, creating an urgent need for effective prediction and prevention strategies. Here, the fusion of GIS and statistical modelling is indispensable in tackling the problem at hand. These tools not only help identify the spatial and temporal patterns of crimes but also aid in formulating data-driven policies and interventions.

The study of crimes against women is critical in countering this societal menace. Gender-based violence is a global epidemic with profound consequences for individuals, families, and societies. In India, and specifically in West Bengal, the challenges related to such crimes are exacerbated by socio-economic and cultural factors that often prove to be stumbling blocks to effective responses and interventions. This book is a timely and vital contribution to help understand and address these challenges through a rigorous and innovative approach.

The state of West Bengal offers a unique vantage point for such a study. Known for its rich cultural heritage and historical significance, West Bengal has seen an upsurge in crimes against women over the past decades. While West Bengal, one of India's most populous states, mirrors overarching challenges at the national level, its distinctive geography in terms of urban-rural divide, cultural ethos, and political dynamics mire the issue in added layers of complexity. Historically, the state has been at the forefront of progressive movements, yet rising crime rates against women paint a contrasting picture.

The primary objective of this book is to harness the power of GIS and statistical modelling to predict crimes against women in West Bengal. By doing so, it aims to provide a scientific framework for developing preventive strategies and policies. The book focuses on four major types of crimes perpetrated against women and girl children: trafficking, rape and sexual harassment, domestic violence, and dowry deaths. A detailed and in-depth study, supported by an extensive literature review and available data on such crimes, both at national and state levels, offers an opportunity to compare and contrast incident patterns at national and state levels.

Each chapter addresses these objectives as each crime has its own distinct spatial pattern, posing differing threats to targets and increasing the risk of victimisation. The study showers light on how women throughout Bengal are exposed to a variety of gender-linked harassment due to situational environmental determinants as well as culturally fostered misogyny, including the persistence of patriarchal values and beliefs, victim-blaming, socio-economic marginalisation, and many other factors responsible for a surge in crimes against women as well as the under-reporting and non-reporting of such incidences.

The integration of GIS and statistical modelling into crime prediction represents a cutting-edge advancement in criminology and public safety. GIS technology, with its ability to map and analyse spatial data, offers invaluable insights into the geographical distribution of crime. When combined with sophisticated statistical models, it provides a powerful tool for identifying patterns, predicting trends, and formulating evidence-based strategies for crime prevention and intervention.

Through a series of well-researched chapters, the authors present a nuanced analysis of the spatial and temporal aspects of crimes against women, offering a rich understanding of the factors contributing to these offences. The use of GIS allows for a visual representation of crime data, which is instrumental in revealing patterns and trends that may not be immediately apparent through traditional statistical methods alone.

The scope of this book extends beyond mere academic inquiry; it is a call to action for policymakers, law enforcement agencies, and social workers. The insights derived from this study are intended to inform and shape policies that ensure the safety and well-being of women. In an era where data-driven decision-making is paramount, this book is a testament to the critical role of research and technology in societal advancement.

The book is more than just an academic endeavour; it is a crucial step towards ensuring the safety and dignity of women. The insights and recommendations outlined in this book have the potential to inform decision making and bring about tangible changes on the ground. This book stands as a beacon of what can be achieved when rigorous research meets state-of-the-art technology.

Department of Geography	Debendra Kumar Nayak
North Eastern Hill University	
Shillong, Meghalaya, India	

Consent Forms

Informed Consent

Informed consent was obtained from all individual participants included in the study.

Ethical Approval

All procedures performed in studies involving human participants were in accordance with the ethical standards of the institutional and/or national research committee and with the 1964 Helsinki Declaration and its later amendments or comparable ethical standards.

Conflict of Interest

The authors declare that they have no conflict of interest.

Permissions

The authors declare that they do not use any figure, table, or lengthy text passage that has been previously published in a copyrighted source. So, no third-party permissions are required.

Acknowledgment

I, author Dr. Priyanka Biswas, wish to extend my sincere gratitude to my coauthor Professor Nilanjana Das Chatterjee, HOD, Department of Geography, Vidyasagar University, for giving valuable supervision, scholarly guidance, and encouragement as well as expanding my vision and thinking during this study. I also treasured her constructive criticisms and valuable suggestions which enriched this work. Her generosity, punctuality, and unwavering support have given me a new direction in my academic career.

We are thankful to the University Grants Commission (UGC), India and Indian Council of Social Science Research (ICSSR, File No. Sc-2/ICSSR/2016-17/RPS) for funding this research work.

We wish to extend our sincere appreciation to all the officials of all departments including the National Crime Records Bureau (NCRB), Open Government Data Platform India, Census office, Development and Planning Department, Government of West Bengal, West Bengal Commission for Women, Ministry of Health and Family Welfare, and other departments of state and central government for providing us valuable information during the intensive study.

We sincerely thank the Vice-Chancellor and Registrar of Vidyasagar University for the administrative support required to conduct this study.

We are grateful to all the faculties of the Department of Geography, Vidyasagar University for their valuable suggestions, cordial support, and encouragement throughout the work. We gratefully acknowledge the support rendered in many ways by all the co-researchers in our department.

We fully express our heartfelt gratitude to our families for their endless support, understanding, and motivation throughout the work and all the happiness, love, and cordial affection they bring to our lives.

Department of Geography, Vidyasagar University	Priyanka Biswas
Midnapore, West Bengal, India	Nilanjana Das Chatterjee

Contents

1	**Statistical Tools for the Spatial Pattern Analysis of Crime Using GIS: An Integrated Approach**................................	1
	Introduction: Scope and Extent....................................	1
	Relevance of West Bengal as the Study Area......................	4
	Methodology..	7
	Datasets...	7
	Data Aggregation and Processing..............................	8
	Analysis Task..	8
	Elaboration of Spatial Statistical Tools and GIS Techniques for Crime Analysis..	9
	Exploratory Data Analysis (EDA)..............................	9
	Spatial Weights Matrix, W.......................................	11
	Estimates of Spatial Dependence: Moran's I................	11
	Spatial Regression Model of Crime Analysis.......................	12
	Geographically Weighted Regression (GWR)................	13
	Assessing Risk Factors Using Statistical Tools....................	14
	Factor Analysis (FA)...	14
	Cronbach Alpha..	15
	KMO (Kaiser-Meyer-Olkin) and Bartlett's Test..............	15
	Relevance of the Study...	16
	References...	17
2	**Brief Elaboration of District-Wise Socioeconomic Settings in West Bengal**...	23
	Introduction...	23
	Administrative Setup of West Bengal...............................	23
	Demography and Socioeconomic Environmental Settings in West Bengal...	24

3	**Geovisualization and Prediction of Crime Against Women in West Bengal Using Statistical Modeling**	33
	Exploratory Spatial Analysis of Crime Data	33
	Method	34
	Periodic Insights into Data: Multivariate Mapping and Space-Time Attribute Visualizations	34
	Geospatial Extent of Crime Concentration: Some Case Studies Based on Newspaper Reports	42
	Results of Spatial Autocorrelation (Moran's I)	44
	Crime Prediction with the ARIMA Model	48
	Model Architecture of ARIMA	49
	Results of ARIMA	50
	Discussions and Conclusion	51
	Appendices	52
	Appendix 1: Summarization of Some Published Newspaper Reports on Women Trafficking in North- and South-24 Parganas Districts, West Bengal	52
	Appendix 2: Summarization of Some Published Newspaper Reports on the Incidences of Acid Attacks Against Women in West Bengal	56
	References	63
4	**Predicting Future Crime Hotspots Using Statistical Techniques and GIS**	65
	Introduction	65
	Understanding Geography of Crime Against Women in West Bengal	66
	Core Risk Factors of Crimes Against Women in West Bengal	68
	Data Normality	69
	Result of Principal Component Analysis	69
	Measures of Collinearity	70
	Spatial Association of Core Risk Indicators: Results of Geographically Weighted Regression (GWR)	71
	Results of GWR	71
	Predicting Future Potential Areas of Crime Against Women in West Bengal	75
	Discussions and Conclusion	75
	Appendix: Local Coefficient Estimations of Variables Based on the GWR Model	77
	References	78
Index		81

List of Figures

Fig. 1.1	Trends of crime against women in West Bengal (2010–2022). (Source: NCRB Report)	4
Fig. 1.2	District-wise spatial intensity of crime against women in West Bengal during 2022. (Source: data.gov.in, 2022)	6
Fig. 1.3	Multidimensional data model: an aggregation of space-time attribute. An illustrative view of a time series for a district and crime type is highlighted	8
Fig. 1.4	Methodological overview	17
Fig. 2.1	District fact sheet of Bardhaman, West Bengal, India	26
Fig. 2.2	District fact sheet of Jalpaiguri, West Bengal, India	27
Fig. 2.3	District fact sheet of Malda, West Bengal, India	28
Fig. 2.4	District fact sheet of Murshidabad, West Bengal, India	29
Fig. 2.5	District fact sheet of North-24 Parganas, West Bengal, India	30
Fig. 2.6	District fact sheet of South-24 Parganas, West Bengal, India	31
Fig. 3.1	Evolvement patterns of cruelties by husband or his relatives (498-A IPC) across space and time in West Bengal. The visualization comprises (**a**) a map-matrix, (**b**) a parallel coordinate plot, and (**c**) a heatmap. (Source: data.gov.in)	35
Fig. 3.2	Evolvement patterns of dowry deaths (304-B IPC) across space and time in West Bengal. The visualization comprises (**a**) a map-matrix, (**b**) a parallel coordinate plot, and (**c**) a heatmap. (Source: data.gov.in)	38
Fig. 3.3	Evolvement patterns of rape (376 IPC) across space and time in West Bengal. The visualization comprises (**a**) a map-matrix, (**b**) a parallel coordinate plot, and (**c**) a heatmap. (Source: data.gov.in)	40
Fig. 3.4	Spatial extent of trafficking hotspots in North- and South-24 Parganas districts. (Source: Newspaper reports (The Hindu, Times of India, The Telegraph, Anandabazar Patrika, Ei-Samay))	43

Fig. 3.5	Spatial extent of incidences of acid attacks against women in West Bengal. (Source: Newspaper reports (Times of India, The Telegraph, The Hindu, Anandabazar Patrika) and ASFI records. (Modified after Biswas & Chatterjee, 2018))	45
Fig. 3.6	Spatial autocorrelation (Moran's I) pattern of different forms of crime against women in West Bengal: (**a**) cruelties by husband or his relatives (sec. 498-A IPC), (**b**) dowry deaths (sec. 304 B IPC), (**c**) rape (sec. 376 IPC), (**d**) kidnaping and abduction (secs. 363–373 IPC), (**e**). overall crime against women. *$p < 0.01$. (Source: Author's computation).	46
Fig. 3.7	Geographical clustering patterns of crime against women in West Bengal. (**a**) Spatial intensity across districts from 2010 to 2022, (**b**) high and low value clusters of districts, p value >0.05. (Source: data.gov.in; author's computation)	47
Fig. 3.8	Architecture of the ARIMA model	50
Fig. 3.9	Forecasting crime against women in West Bengal using ARIMA (1, 3, 4)	51
Fig. 4.1	The standard residuals of geographically weighted regression	73
Fig. 4.2	Local coefficient estimations for factors of crime against women in the GWR model: (**a**) PC1, (**b**) PC2, (**c**) PC3, (**d**) PC4	74
Fig. 4.3	Future potential vulnerable areas of crime against women in West Bengal	75

List of Tables

Table 1.1	Most convenient EDA techniques based on the type of data and the objectives of the analysis	10
Table 1.2	Consistency range of alpha (α) value	15
Table 2.1	Administrative divisions of West Bengal	24
Table 3.1	Model statistics	51
Table 4.1	KMO and Bartlett's test result	69
Table 4.2	Factor loading of determinants of crime against women in West Bengal	70
Table 4.3	Collinearity statistics	71
Table 4.4	Results of the GWR model	72

About the Authors

Priyanka Biswas completed her Ph.D. from the Department of Geography, Vidyasagar University, West Bengal, India. Her areas of research interest are criminological studies especially crime against women, urban environment and crime, and statistical modeling. She emphasizes multidisciplinary efforts to visualize the crime scenarios and evaluate policy-level change to make society crime-free.

Nilanjana Das Chatterjee is a professor and HoD in the Department of Geography, Vidyasagar University, Midnapore, West Bengal, India. Her principal areas of research interest are environmental studies, criminological studies, criminal psychology, and associated socio-economic milieu. She emphasizes to use evaluation strategies to develop community-level awareness and support policy-level change to enhance the quality of society.

Chapter 1
Statistical Tools for the Spatial Pattern Analysis of Crime Using GIS: An Integrated Approach

Introduction: Scope and Extent

To have a better understanding of crime morphologies, modern criminological literature stresses the growing need to understand spatial patterns in order to aid in the development of proactive policing, the delineation of predictive hotspots, and the forecasting of criminal activities (Gorr et al., 2003; Johnson & Bowers, 2004; Cohen et al., 2004; Wu & Grubesic, 2010; Anselin et al., 2000). Yet a complete understanding of criminogenic phenomena requires a consideration of a dynamic range of analysis, including the detection of clusters and the mapping of crime hotspots (Craglia et al., 2000; Murray et al., 2001; Chainey & Ratcliffe, 2005; Eck et al., 2005; Chainey et al., 2008; Wu & Grubesic, 2010); understanding the underlying backcloth, i.e., the physical and socioeconomic environment that fuel the emergence of crime concentrations in a space (Brantingham & Brantingham, 1981; Morenoff et al., 2001; Gorman et al., 2001); understanding the theoretical context, i.e., how a specific space provides opportunities for crime and influences crime patterns (Shaw & McKay, 1942; Morenoff et al., 2001; Cohen & Felson, 1979; Messner & Anselin, 2004); developing improved methods and techniques to better analyze crime data (Anselin et al., 2000; Bernasco & Elffers, 2010; Levine, 2006); and establishing effective models for predicting crime hotspots and suggestive measures for proactive policing (Hunt et al., 2008; Ratcliffe, 2004a). Regardless of the understanding of the underlying backcloth, criminologists recognize the importance of detecting crime hotspots, analyzing distributional patterns, and intervening in crime-ridden areas (Braga, 2001; Ratcliffe, 2004b). Not only the criminological literature but also criminogenic studies draw substantial attention from a geographical perspective. Since the 1830s, the geographical perspective has shown considerable attention to understanding the occurrence of criminogenic events and the elaboration of crime prevention measures (Guerry, 1833; Harries, 1974; Weisburd et al., 2009; Liu & Eck, 2008; LeBeau & Leitner, 2011). The geography of crime stresses the importance of unraveling the underlying backcloth of criminogenic situations to enable effective space-based intervention policing while bolstering community safety. Practitioners in geography consider "space" as a significant attribute to gain deep insights into crime patterns and implement targeted interventions. From a spatial perspective, geographic studies on crime have revealed that instances of criminogenic events are not distributed randomly over space (Sherman et al., 1989; Roy & Chowdhury, 2023a, b; Chainey & Ratcliffe, 2005) but are clustered in small discrete geographical areas (Ratcliffe, 2004a, b; He et al., 2017; Sherman & Weisburd, 1995). Moreover, empirical shreds of evidence have demonstrated the importance of space and indicate that an area's specific socioeconomic and environmental attributes can significantly contribute to making that area a hotbed of crime (Roy & Chowdhury, 2023a, b; Santos, 2016). Besides, the importance of a space-based study on criminogenic events has gained universal attention, specifically in the Western world (Weisburd et al., 2010). Therefore, the spatial pattern analysis of crime has evolved significantly in criminology, geography, as well as related socioeconomic planning sciences over the past decade (Chainey et al., 2008; Eck et al., 2005). In addition to understanding the spatial distributional patterns of criminogenic events, a consideration of the temporal trends in crimes at varying temporal dimensions is also crucial among criminologists and crime prevention practitioners (Rengert, 1997; Townsley et al., 2000; Henry & Bryan, 2000; Felson & Poulsen, 2003; Weisburd et al., 2004, 2009; Townsley, 2008). Although in criminogenic studies both space and time are considered fundamental attributes, to facilitate effective and legitimate crime prevention and intervention measures, the consideration of both space and time plays a decisive role in understanding crime dynamics (Roy & Chowdhury, 2023a, b). Criminologists reveal that the temporal evolution of spatial crime patterns needs to be understood before implementing evidence-based prevention measures to address criminogenic nuisance effectively (Roy & Chowdhury, 2023a, b; Jiang et al., 2021). Because stable and unstable crime hotspots may exist simultaneously in a particular space (Johnson & Bowers, 2004, 2008), if any reduction happens in the spatiotemporal scale of measures, crime hotspots become more dynamic and unstable (Mohler et al., 2017). Thus, to avoid the wastage of scarce resources and enhance crime deterrence effects, criminologists and other practitioners undertake intervention measures in comparatively stable crime hotspots rather than in dynamic or unstable ones. Therefore, in space-based criminogenic studies, a consideration of the spatial and temporal evolution of crime is crucial among practitioners to make effective prevention measures that focus on the availability of resources and implement targeted interventions. Another important facet of the geography of crime is understanding the context and the underlying socioeconomic processes that eventually affect the spatial and temporal variations of crime (Hagenauer et al., 2011). Grasping an understanding of the spatial dependencies and socioeconomic predictors of crime is pivotal for yielding valuable insights into crime dynamics and determining geographical areas that necessitate efficacious targeted intervention and resource allocation to address crime (Kabiraj, 2023; Roy & Chowdhury, 2023a, b). For this widespread understanding of criminogenic events in an area, the application of geostatistical tools is pivotal for visualizing and modeling crime. Henceforth, the integration of GIS (geographical information system) and statistical methods has become popular among practitioners in recent days. GIS provides practitioners with a powerful set of tools for visualizing and analyzing spatial data, which aids in legitimate crime prevention (Krivoruchko et al., 2003). It emerged as a key tool in intelligence-led policing to spatially predict criminogenic events and assist patrol units in making cost-effective crime prevention decisions (Ratcliffe, 2012; Fitterer et al., 2015). Considering the geographical context of crime, which incorporates both space and time, map-based spatiotemporal analysis in GIS can help patrol units comprehend real-time information on the patterns of crime, aiding in decision-making processes (Ratcliffe, 2012; Chainey & Ratcliffe, 2005; Wang, 2012). A geostatistical analyst conveyed that these GIS tools can be even more powerful when integrated into statistical methods for spatial data analysis (Krivoruchko et al., 2003). The application of geostatistical tools and techniques in criminogenic understanding can be beneficial to unraveling the underlying occurrence mechanism of crime and predicting the likelihood of crime occurrence, which may lead to decision-making processes. Grasping the understanding of crime patterns, spatial dependencies and socioeconomic and environmental predictors of crime, application of geovisual and geostatistical tools is pivotal. A geovisual analytic tool allows for the simultaneous visualization of multiple dimensions and the discovery of interesting information through a variety of combined perspectives. The geostatistical models of spatial analysis on the other hand can provide a powerful set of interactive, analytical tools uniquely suited to the goal of mapping, modeling, and predicting crime. Henceforth, the major objectives of this present work are to systematically geovisualize and analyze the complex patterns of crime against women in the context of West Bengal, a segment of the Indian subcontinent; to apply suitable geostatistical analytic tools to unravel the occurrence mechanisms underlying crime patterns; and to conduct geostatistical modeling of crime to predict potentially vulnerable areas for policy purposes. The main goal is to meticulously geovisualize and analyze various crime patterns and develop a standardized model for systematically predicting crime and geospatially analyzing criminogenic events, thus aiding in intelligence-led policing to prevent and control crime. The proposed workflow is as follows: Chap. 1 provides a brief overview of the study area, describes available gross data sources, outlines the variables of interest considered for analyzing facts, elaborates on integrated data-mining techniques for exploratory spatial data analysis, and discusses the amplification of GIS tools and statistical modeling for geospatial analysis and crime prediction. Chapter 2 briefly presents the district-wise socioeconomic fact sheets of West Bengal. Chapter 3 discusses periodic insights into crime data, specifically the space-time visualization of different crime patterns and the predictive analysis of crime scenarios against women in West Bengal, using relevant statistical modeling. Exploratory spatial data analysis, the determination of crime hotspots, spatial autocorrelation, regression results, and crime prediction and discussion are also covered in this chapter. Next, Chap. 4 presents the core risk indicators of crime against women in West Bengal, elaborates on the spatial association of potential factors and occurrence mechanisms, and predicts future crime hotspots using statistical modeling by incorporating spatially contextual factors within a GIS environment. Finally, this study concludes with a robust discussion of the study findings and future scope of this work. The entire work has been structured meaningfully to provide all relevant facts while maintaining the continuity of the research work. This work possesses a noteworthy understanding of the criminogenic nuisance against women in West Bengal, which is pivotal for adopting efficacious intervention strategies and policies to prevent and address crime.

Relevance of West Bengal as the Study Area

The eastern Indian state of West Bengal has been experiencing gender-related atrocities for a long time. From the "Sati" period to the modern era, gender-related atrocities in West Bengal have been deeply rooted in the patriarchal culture. Over the past decade, gender-based atrocities against women in West Bengal have reached a maximum, which raised big concerns among practitioners, policymakers, criminologists, and social activists about the safety of women in Bengal. This poses a great challenge to practitioners to take the best possible measures to address crimes against women in West Bengal and provide a safer environment so that women may flourish with all their inner possibilities. As per the recently published National Crime Records Bureau (NCRB) report, West Bengal ranks fourth in terms of crimes committed against women in India, with 34,738 numbers of reported cases (IPC + SLL), after Uttar Pradesh (65,743 cases), Maharashtra (45,331 cases), and Rajasthan (45,058 cases) (NCRB, 2022). Considering the last 12 years of NCRB reports, it can be observed that since 2010, incidences and rates of crime against women in West Bengal have been following a continuous upward trend (Fig. 1.1). In 2010, the rate of crime against women in West Bengal was 29%, which reached its maximum in 2014 (85.4%), and after that, it continued to follow a parallel rising trend.

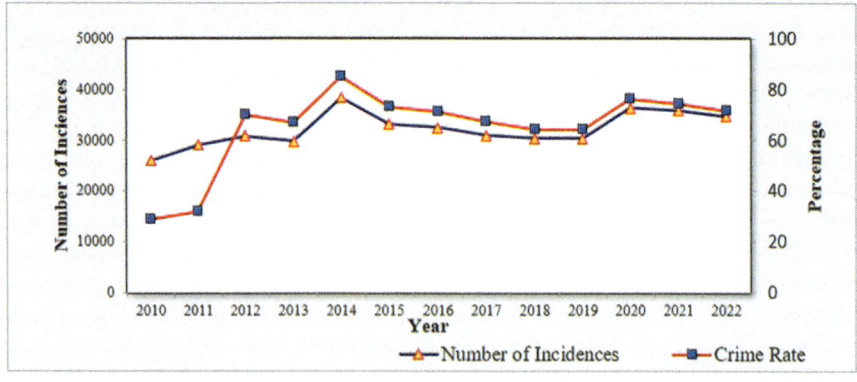

Fig. 1.1 Trends of crime against women in West Bengal (2010–2022). (Source: NCRB Report)

Apart from the overall scenarios of crime against women, significant spatial variability has been observed throughout West Bengal concerning the ferocity that women face in their everyday lives. As per the 2022 district-wise open government data (data.gov.in) provided by the Ministry of Home Affairs, crime against women in West Bengal is mostly predominant in the districts of South-24 Parganas, Murshidabad, North-24 Parganas, Bardhaman, Nadia, Purba Medinipur, Malda, Howrah, Kolkata, Hooghly, and Cooch Behar (Fig. 1.2). Among the metropolitan cities in India, Kolkata, the state capital of West Bengal, is in a vulnerable position (crime rate of 27.8) regarding the incidences of crimes committed against women (NCRB, 2022). Not only that, but spatial variability has also been observed concerning the specificity of crimes. Crimes against women are manifested in several forms in West Bengal. They range from sex-selective abortion, infanticide, child abuse, child marriage, domestic abuse, dowry fatality, cruelties by husbands and their relatives, honor killings, rape, kidnapping, forced prostitution, women trafficking, acid attacks, and many other forms of beastliness against women. Each crime has its distinct spatial patterns, posing differential threats to targets and exacerbating the risk of victimization. According to the 2022 district-wise open government data (data.gov.in), North-24 Parganas ranks highest (3054 cases) when it comes to cruelties by husbands or their relatives (498-A IPC), followed by South-24 Parganas (2825 cases), Murshidabad (2002 cases), and Bardhaman (1331 cases). For incidences of dowry deaths (304-B IPC), the district of North-24 Parganas reported the highest number of cases (61) during 2022, followed by South-24 Parganas (43) and Murshidabad (41). In terms of rape (376 IPC), the district of Malda reported the highest number of cases (187), followed by North-24 Parganas (144), Murshidabad (118), and Nadia (91). The border districts of West Bengal, viz., North-24 and South-24 Parganas, Malda, Nadia, Murshidabad, Cooch Behar, Uttar and Dakshin Dinajpur, and Jalpaiguri, have emerged as hotbeds for illicit activities such as trafficking, continuously putting women at high risk of threat. The porous borderland location, enormous illegal immigration, and adverse socioeconomic conditions shape a situational environment of illicit activities in these borderland areas. Existing studies also indicate that adverse physical environments and the resulting socioeconomic vulnerabilities in some regions, such as remote villages in the deltaic Sundarbans of South-24 Parganas, remote hilly villages, and tea garden areas of the Darjeeling Himalayas, create conditions favorable for the trafficking of women and minors (Chakrabarti & Sarkar, 2007; Ghosh, 2014; Biswas and Das Chatterjee, 2024; Molinari, 2017; Biswas and Chatterjee, 2021). It has evolved, and specific types of crimes against women are now spatially concentrated in specific pockets of West Bengal, forming distinct spatial patterns. Space-specific factors, along with culturally fostered misogyny like patriarchal ideology, gender norms, victim-blaming attitudes, socioeconomic marginalization, and associated factors, have resulted in the breaking of social ties, heightened insecurities among women, and an increased rate of violence victimization. These spatial variabilities bring difficulties in the implementation of proper measures to control violence victimization among women at large. Hence, the issue of women's safety in West Bengal has become a matter of great concern among practitioners in recent days.

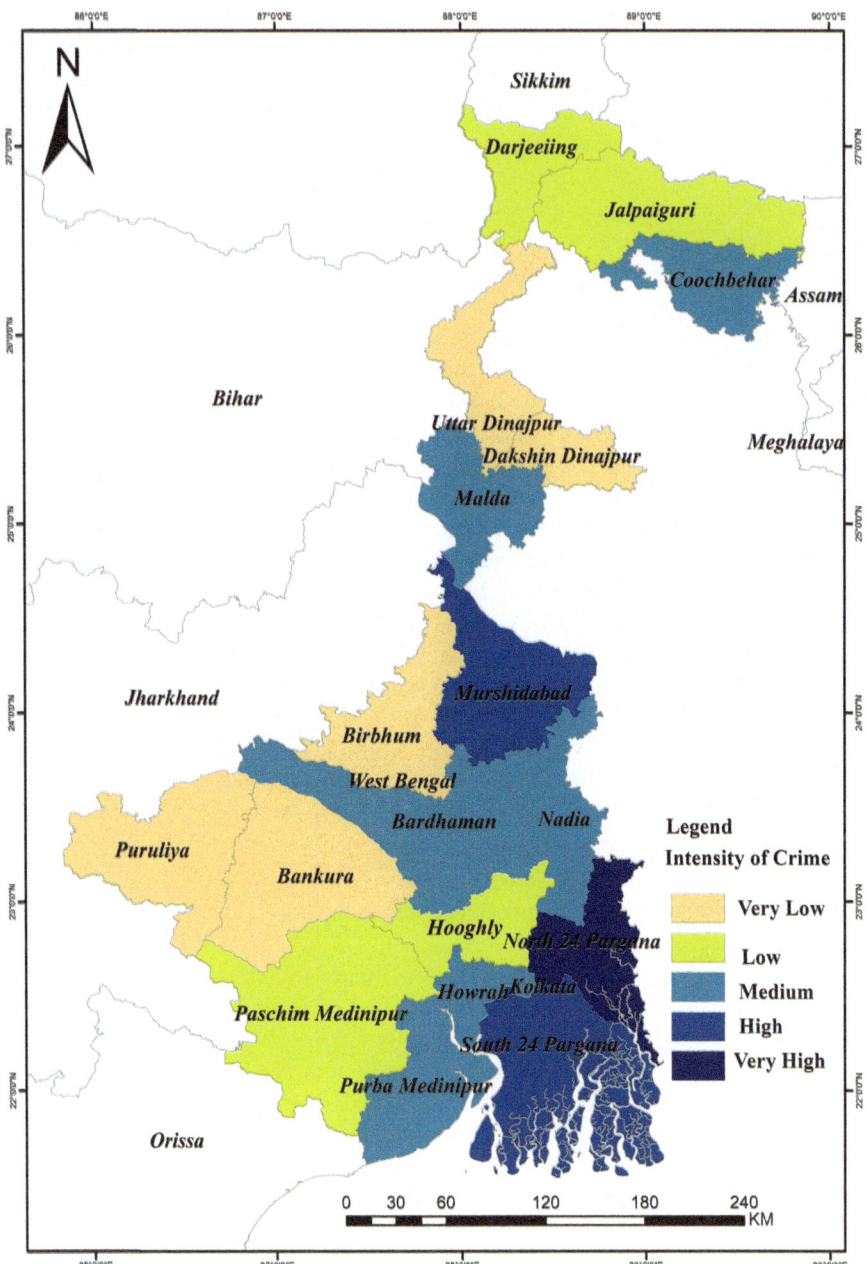

Fig. 1.2 District-wise spatial intensity of crime against women in West Bengal during 2022. (Source: data.gov.in, 2022)

Methodology

Datasets

The datasets used for this study have been collected from the official web portal of data.gov.in, which provides district-wise crime data across India. To focus on the basic tenets of geovisualization and the analysis of criminogenic phenomena from a geographic standpoint, four major crime categories from 2010 to 2022—cruelties by husband or his relatives (Sec. 498-A IPC), dowry deaths (Sec. 304-B IPC), rape (Sec. 376 IPC), and kidnapping and abduction (Secs. 363–373 IPC)—have been considered for intensive study based on the evaluation of the most-reported official records. Also, gross crime data during that time span (2010 to 2022) has been considered to analyze crimes across West Bengal. Due to the unavailability of geocoded datasets, as crime information is only recorded at a more aggregate level across census block groups, 18 districts of West Bengal have been considered as spatial units, along with temporal variations and crime types. It is to be noted that the newly formed districts in West Bengal, namely, Kalimpong, Alipurduar, Paschim Bardhaman, and Jhargram, have not been included in this study since most of the secondary information used to analyze the spatial association of crime is based on the 2011 census data, and the most recent census data is not available yet. There are four additional reasons for considering districts (or neighborhoods) as base units in this study. First, at this level of spatial extent, broader patterns of neighborhood distress and spatial heterogeneity might become more obvious. A comprehensive understanding of differential crime patterns across multiple perspectives might be apparent at this level. Second, each district (or say neighborhood) has a definite geographical extension that directly supports efficacious place-based policing and resource allocation. Third, from a broad perspective, it might be helpful to determine the overall pertinent factors responsible for different crime patterns, as most of the socioeconomic databases are available at the district level only. The socioeconomic attributes considered for determining the inherent factors accountable for women susceptibilities throughout West Bengal include district-wise population density, sex ratio, literacy rate, number of nonworking population, number of immigrants, decadal growth rate (data source: Census of India, 2011), GDP (gross domestic product), HDI (Human Development Index) (data source: Department of Statistics and Programme Implementation, WB 2014–2015), HPI (Human Poverty Index) (data source: District Statistical Handbook), amenities (data source: Economic Review 2017–2018 published by the Department of Planning, Statistics and Programme Monitoring, West Bengal), and school enrollment ratio (District Statistical Handbook, 2011). These attributes have been selected after an in-depth review of the existing literature on the pertinent factors related to crime against women in our society.

Data Aggregation and Processing

The available datasets under different crime categories are aggregated and arranged in separate attribute tables and transformed depending on the analysis task. In each attribute table, the row denotes the spatial dimension (say district), and the column indicates the attribute value (e.g., crime) at a temporal extent. Each cell value has been represented as the total number of crimes happening in a particular spatial unit (neighborhood) at a specific time frame, for example, the total number of kidnapping and abduction cases reported in Murshidabad district in the year 2015. A three-dimensional illustrative view of the attribute table is represented in Fig. 1.3. Once the formation of the attribute table is completed, as illustrated above, it has been normalized or standardized depending on the analysis task, which is described below.

Analysis Task

To understand how the temporal evolution of crime patterns varies across space and crime types, the crime count in each cell in a particular time series has been divided by the total crime count of that time series. Henceforth, each cell value was

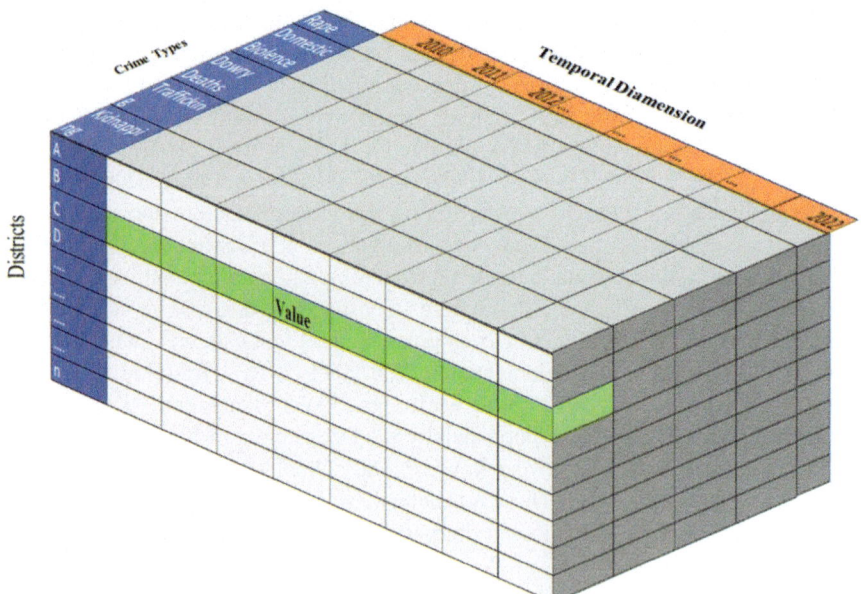

Fig. 1.3 Multidimensional data model: an aggregation of space-time attribute. An illustrative view of a time series for a district and crime type is highlighted

converted to a percentage value, representing the proportion of crimes for each time period for a particular neighborhood (district) and crime type.

Elaboration of Spatial Statistical Tools and GIS Techniques for Crime Analysis

Exploratory Data Analysis (EDA)

Exploratory data analysis (EDA) is a key step after the collection and preprocessing of datasets to discover patterns in the data (Komorowski et al., 2016). The primary aim of EDA is to assist the analyst in delving into given datasets and examining the data quality and distribution patterns, detecting outliers and anomalies, and developing models (Komorowski et al., 2016). Since Tukey's seminal work in 1977, EDA has gained a large following as a state-of-the-art technique for dataset analysis (Mosteller & Tukey, 1977; Tukey, 1977). Seltman (2012) explicitly defined the term "exploratory data analysis" (EDA) as a method for gaining insight into data without incorporating formal statistical modeling and inference. It provides conceptual and computational tools for hypothesis development and refinement after visualizing and understanding the available datasets, usually through graphical representation. Many of the EDA techniques used in the analysis of datasets are also quantitative in nature. Since the basic purpose of EDA is to explore and the role of a data analyst is to listen to as much data as possible until a reasonable "story" of the data is evident, graphics provide the analysts with unparalleled power to do so while giving insight into the data (Behrens, 1997; Komorowski et al., 2016). There are many EDA techniques that exist for the identification of significant patterns of any data. Therefore, in a broad sense, EDA methods can be cross-classified in two ways: (1) graphical or nongraphical and (2) univariate or multivariate (mostly bivariate) (Komorowski et al., 2016; Seltman, 2018). Graphical methods visualize the data diagrammatically, whereas nongraphical methods summarize the data using statistical calculation. Univariate methods incorporate one variable at once, whereas multivariate methods consider two or more variables to explore the potential relationships between the variables. Univariate nongraphical EDA techniques include measures such as central tendency, spread, standard deviation, skewness, kurtosis, and outliers. On the other hand, univariate graphical techniques include histograms, stem-and-leaf plots, boxplots, quantile-normal (QN) plots, and quantile-quantile (QQ) plots. Multivariate nongraphical EDA techniques usually incorporate either cross-tabulation (bivariate nongraphical EDA) or statistics (correlation and covariance) to show the relationship between two or more variables. And the most admissible graphical multivariate EDA techniques incorporate side-by-side box plots, scatterplots, etc. Beyond the above-mentioned cross-classification, EDA techniques can further be categorized depending on the type of data and the objectives of the analysis. In this context, Komorowski et al. (2016), in their study, summarized the most convenient EDA techniques, which are illustrated in Table 1.1. Here, it should be mentioned that although EDA is an important art for analyzing any data pattern, performing appropriate EDA is always necessary to become more acquainted with the data, avoid obvious mistakes, and learn about variable distributions and relationships between the variables.

In the case of spatial pattern detection for the determination of areas exhibiting heightened susceptibilities to specific phenomena such as "crime against women," practitioners consider exploratory spatial data analysis (ESDA) as a significant segment of EDA. This analytical process entails the adoption of a state-of-the-art technique that aims to detect hotspots or areas characterized by heightened levels of specific phenomena (e.g., criminogenic nuisance) (Tan & Haining, 2009; Messner et al., 2013; Zakaria & Rehman, 2015). It is a powerful tool that facilitates analysts in exploring complex spatial features, identifying clusters and spatial outliers, detecting patterns of spatial association, etc., within a given dataset (Kumar et al., 2012; Roy & Chowdhury, 2023a, b). Thus, the selection of suitable techniques to carry out spatial pattern detection is typically viewed as exploratory spatial data analysis (ESDA) (Anselin, 1998; Murray & Estivill-Castro, 1998; Wu & Grubesic, 2010), yet in some contexts, it may also be confirmatory (Gelman, 2004). An integral aspect intrinsic to ESDA that pertains to exploring spatial patterns of criminogenic events is the integration of GIS and crime analysis. Therefore, spatial autocorrelation, kernel density mapping, and cluster analysis are some popular practices for the detection of spatial crime patterns (Majumder et al., 2023; Shafeeq et al., 2014; Messner et al., 1999; Anselin et al., 2009). Nevertheless, Grubesic (2006) considers cluster analysis to be relatively underutilized in crime pattern detection due to confusion regarding the selection of hierarchical and nonhierarchical methods, the emergence of spatial biases related to geographic space, and limited guidance on the determination of the exact number of clusters. For spatial

Table 1.1 Most convenient EDA techniques based on the type of data and the objectives of the analysis

Data type	Suggested EDA techniques	Objectives	Suggested EDA techniques
Univariate continuous	Histogram, line plots	Understanding distribution patterns	Histogram
Bivariate continuous	2D scatter plots	Determination of outliers	Histogram, box plots, scatter plots
Categorical	Central tendency, standard deviation, skewness, kurtosis, spread, outliers	Quantifying the relationship between variables (one exposure and one outcome)	Scatter plots, curve fitting, covariance, and correlation
Multiple groups	Side-by-side box plots	Visualizing the relationship between variables	Heatmap, PCP
2D arrays	Heatmap	Visualization of high-dimensional data	2D/3D scatter plots, PCA

Source: Komorowski et al. (2016); modified by author
Note: *PCP* parallel coordinate plot, *PCA* principal component analysis

pattern analysis, spatial autocorrelation is considered a powerful technique that quantifies the degree of similarity or dissimilarity between proximate locations (Cliff & Ord, 1981; Shafeeq et al., 2014; Roy & Chowdhury, 2023a, b). It functions as an important index to evaluate spatial dependency and the heterogeneity of variables, i.e., how values attributed to a variable at a certain location correlate with those in neighboring areas, blocks, or districts (Roy & Chowdhury, 2023a, b). It rests on the principle that geographically contiguous sites are more likely to exhibit similar values owing to underlying processes or factors (Anselin et al., 2007; Zakaria & Rehman 2015; Dou et al., 2016; MacIndoe & Oakley, 2023). Several statistical measures can be applied to quantify spatial autocorrelation, including Moran's I, local indicators of spatial association (LISA), Gettis-Ord Gi*, and spatial clustering (Shafeeq et al., 2014; Roy & Chowdhury, 2023a, b). These measures assess the magnitude and statistical significance of spatial linkages between observations. They are instrumental in determining clusters within geographic space (Roy & Chowdhury, 2023a, b). Furthermore, these measures help specify spatial patterns or trends while appraising the presence of spatial associations.

Spatial Weights Matrix, W

The spatial weights matrix, *W*, is a prerequisite for ESDA. It is also referred to as the "Cliff-Ord weight matrix" (Anselin, 2013). The determination of the appropriate spatial weights for the elements of this matrix is crucial in ESDA (Anselin, 1998, 2013). Therefore, to ascertain the matrix accurately, the spatial weights matrix *W* is calculated as follows:

$$\begin{bmatrix} W_{11} & W_{12} & \cdots & W_{1n} \\ W_{21} & W_{22} & \cdots & W_{2n} \\ W_{31} & W_{32} & \cdots & W_{3n} \\ \vdots & \vdots & \vdots & \vdots \\ W_{n1} & W_{n2} & \cdots & W_{nn} \end{bmatrix}$$

where *n* represents the number of spatial units (region). If regions *i* and *j* are proximate to each other (neighbors), then $Wij = 1$; otherwise, $Wij = 0$.

Estimates of Spatial Dependence: Moran's I

Moran's *I* stands as a classic measure for spatial autocorrelation employed to evaluate spatial dependence and the clustering tendencies of a variable within distinctive geographical areas. It can be applied to both points and polygons, which have attached attribute data (Chainey & Ratcliffe, 2005). This measure statistically quantifies the extent of similarity between neighboring sites based on values attributed to a specific variable of interest, thus allowing analysts to determine if any significant clustering exists in, for example, a rape pattern distribution in a region (Chainey & Ratcliffe, 2005; Anselin et al., 2007; Majumder et al., 2023). By assessing the spatial association among the observations, Moran's I effectively discerns whether similar values tend to exhibit aggregation or dispersion across geographical space, accordingly revealing positive or negative spatial autocorrelation, respectively. The range of Moran's I varies from -1 to $+1$. On a given significant level, a value closer to 1 is indicative of positive spatial autocorrelation, while a value nearing -1 denotes pronounced negative spatial autocorrelation. If Moran's I value nearer to 0 signifies no or absence of spatial autocorrelation, a manifestation akin to randomness (Chen, 2013). The mathematical expression of Moran's I is as follows:

$$\text{Moran's } I \frac{n}{S_o} \times \frac{\Sigma_i \Sigma_j W_{ij}\left(x_i - \bar{X}\right)\left[x_j - \bar{X}\right]}{\Sigma_i \left[x_i - \bar{X}\right]^2}$$

where n represents the total features count, x denotes the interest variable, W_{ij} signifies the standardized weight matrix between i and j, and S_o denotes the aggregate sum of all spatial weights.

The primary objective of executing Moran's I in this present study is to examine the existence of spatial autocorrelation within a particular set of crimes against women across different districts in West Bengal. The aim is to determine the emergence of any significant discernible spatial pattern, thereby assisting practitioners in assessing whether these criminogenic nuisances exhibit noteworthy geographic clustering or randomly disperse throughout the spatial extent. This analytical work might shed light upon observed patterns, thereby extending our understanding of the crime phenomena.

The initial data exploration stage is crucial as it allows analysts to validate some of the hypotheses (da Silva et al., 2020), like which neighborhoods are highly vulnerable to crimes against women, what types of crimes occur more in those areas, etc. Based on these prior validations, further analysis, data modeling, and future crime prediction could be easier to execute.

Spatial Regression Model of Crime Analysis

Crime mapping is necessary not just for detecting crime hotspots or cold spots but also for determining the underlying drivers that potentially act as catalysts for crime. Consequently, criminologists expend considerable effort in understanding the impact of physical and socioeconomic environmental determinants, such as unemployment, poverty, literacy, and physical environmental settings, to better understand what potentially drives criminal activities. Thus, by influencing specific underlying drivers, it might be possible to address crimes. Several inferential regression analyses have become popular among criminologists in this regard, one of which is geographically weighted regression (GWR).

Geographically Weighted Regression (GWR)

Understanding crime morphology has long been a growing concern among policymakers at the national and regional levels as emerging criminogenic activities in an area erode the social and economic environment, facilitating high-scale disruption, hardship, and adversities and creating large-scale regional crises (GOSCHIN, 2019). These socioeconomic hardships caused by emerging criminogenic nuisances raise multiple challenges for policymakers aiming to safeguard the public. Thus, in light of these imminent difficulties, law enforcement officials and policymakers stresses the importance of determining the underlying backcloths (situational factors) of crime after the initial exploration of crime data. Yet criminologists convey that major situational factors for criminogenic nuisances may exert differential effects at the national and local levels, depending on specific socioeconomic conditions (GOSCHIN, 2019). Besides, obvious territorial correlations regarding crime determinants may emerge at the local level as criminality in a region depends not only on adversities within that region but also on those in nearby ones (GOSCHIN, 2019). Specific spatial econometric procedures account for such spatial correlations by incorporating the use of spatial weights, which signify the impact of each region on its neighboring ones. Therefore, appropriate spatial statistical tools need to be adopted to specify the spatial heterogeneity of criminogenic activities within a microregional context. Among the spatial methods, geographically weighted regression (GWR) is considered a valuable tool for estimating local coefficients specific to each region (Wheeler & Tiefelsdorf, 2005; Fotheringham et al., 2009; Lu et al., 2014) and specifying which regions are more susceptible to specific factors of crime as per spatially defined weights. As classic regression models estimate global coefficients that cannot define significant spatial variation, GWR allows the model to vacillate regionally, providing a better portrait of the spatial variation of the criminogenic phenomenon. Thus, GWR provides useful information for designing appropriate anticrime measures at the regional level, depending on local particularities. Hence, this study applies a criminality GWR model to find local determining factors (both socioeconomic and demographic) that explain the spatial extent of women's susceptibilities to crime in West Bengal. The mathematical expression of a basic GWR model is as follows:

$$y_i = \beta_{i0} + \sum_{k=1}^{m} \beta_{ik} X_{ik} + \varepsilon_i \qquad (1.1)$$

where y_i represents the dependent variable at location i, β_{i0} denotes the intercept at location i, m denotes the number of independent variables, represents the kth independent variable at location i, β_{ik} denotes the local regression coefficient for the kth independent variable at location i, and signifies the random error at location i.

GWR estimates the inherent interrelationships around each regression point i, where each set of regression coefficients is calculated by weighted least squares. The equational matrix for this estimation is as follows:

$$\beta i = \left(S^T W_i S\right)^{-1} S^T W_{iy} \tag{1.2}$$

where S denotes the matrix of the independent variables with column 1 s for the intercept, y denotes the vector of the dependent variable, $\beta i = \left(\beta_{i_0}, \ldots \ldots \beta_{i_m}\right)^T$ is the vector of $m + 1$ local regression coefficients, and W_i represents the diagonal weight matrix specific to location i.

Here, the weights, W_i, are calculated with a kernel function based on proximities between regression points i and the n data points around it. Arc GIS 10.0 software has been used to perform GWR to captured spatially variations of determinants of crime against women in West Bengal from a regional perspective.

Assessing Risk Factors Using Statistical Tools

The application of statistical tools is very much needed to have a proper understanding of the interrelationship among the set of observed variables and determine the most significant risk factors that create the situational backcloth of deviant behavior. An overview of the statistical tools used in the present study is discussed below.

Factor Analysis (FA)

Factor analysis (FA) is a dimension reduction technique that helps determine the interrelationship among variables and also elucidates these variables in terms of the most determining dimensions or factors. This statistical technique helps identify the most significant factors among a range of observed variables with maximum factor loadings. As per Brown (2006), this statistical multivariate procedure helps in executing the underlying factor structure that is highly accountable for facilitating variation and covariation among a set of observed variables. In factor analysis (FA), the present study has followed the exploratory factor analysis (EFA) method using the extraction method of principal component analysis with varimax rotation (Anderson and Gerbing, 1988) to determine the most noteworthy factors accountable for the occurrence of specific criminogenic activities that explain maximum variables in the best possible way. It is worth mentioning that EFA, a data-driven technique, has been used in this study instead of CFA (confirmatory factor analysis) as no existing theory or empirical study has been done before regarding criminogenic issues (Suhr, 2006). KMO (Kaiser-Meyer-Olkin) and Bartlett's test were performed while running EFA to examine data eligibility by measuring the adequacy of the sample for each variable.

Cronbach Alpha

Cronbach alpha is a reliability measurement technique to measure the internal consistency of a scale. It was named after Lee Cronbach in 1951. This is the ratio of true score variance to observed score variance. The value of alpha varies between 0 and 1. Closer to 1 reflects greater consistency among the variables. Reliability is considered an acceptable level if the value is equal to or greater than 0.7 (Nunnally, 1994). The alpha scale is represented in Table 1.2. The equation of Cronbach alpha, a, is as follows:

$$\alpha = \frac{N.\tilde{c}}{\tilde{v}+(N-1).\tilde{c}}$$

Where

N = number of items.
$\tilde{c}$ = average inter-item covariance among the items
$\tilde{v}$ = average variance.

KMO (Kaiser-Meyer-Olkin) and Bartlett's Test

KMO (Kaiser-Meyer-Olkin) and Bartlett's test of sphericity measure the sampling adequacy for each variable, which denotes how suited the data is for factor analysis (Cerny & Kaiser, 1977). The KMO value lies between 0 and 1. To indicate multivariate normality among variables, the worldwide accepted KMO value should be

Table 1.2 Consistency range of alpha (α) value

Cronbach's alpha	Internal consistency
$0.9 \leq a$	Excellent
$0.8 \leq a < 0.9$	Good
$0.7 \leq a < 0.8$	Acceptable
$0.6 \leq a < 0.7$	Questionable
$0.5 \leq a < 0.6$	Poor
$a < 0.5$	Unacceptable

above 0.6, and Bartlett's test of sphericity must be below 0.05. The equation of the KMO test is:

$$\text{KMO}_j = \frac{\sum_{i \neq j} r_{ij}^2}{\sum_{i \neq j} r_{ij}^2 + \sum_{i \neq j} u}$$

where

$R = [r_{ij}]$ is the correlation matrix.
$U = [u_{ij}]$ is partial covariance matrix.

The equation of Bartlett's test is:

$$x^2 = -\left(n - 1 - \frac{2p+5}{6}\right) * \ln R$$

where

p = the number of variables
k = the number of components
I_j = jth eigenvalue of S
$df = (p-1)(p-2)/2$

A glimpse of the general methodological overview of the entire study is presented through a flow chart shown in Fig. 1.4. A detailed chapter-wise elaboration of the methodologies to reach each objective and analysis is discussed later.

Relevance of the Study

The application of geostatistical models in the geospatial understanding of criminogenic consent is becoming pivotal among criminologists in the twenty-first century to aid in intelligence-led policing and crime reduction. The spatial understanding of crime data using geostatistical tools within a GIS environment is crucial for yielding valuable insights into crime dynamics and assisting criminologists in modeling and predicting crime and determining spatial units that require competent cost-effective targeted intervention. The present study meticulously geovisualizes and analyzes multiple dimensions of crimes perpetrated against women in West Bengal, systematically unraveling the underlying occurrence mechanisms and developing a standardized model using geostatistical tools to predict crime. This analytical study sheds light on observed patterns of crime phenomena, extending our in-depth understanding of the spatial extent of crimes, and emphasizes applying spatial statistical tools to postulate the spatial heterogeneity of crimes within a microregional environment based on spatially defined weights. Thus, it would help policymakers detect predictive crime hotbeds in West Bengal, invent proactive space-based policing, and forecast crime.

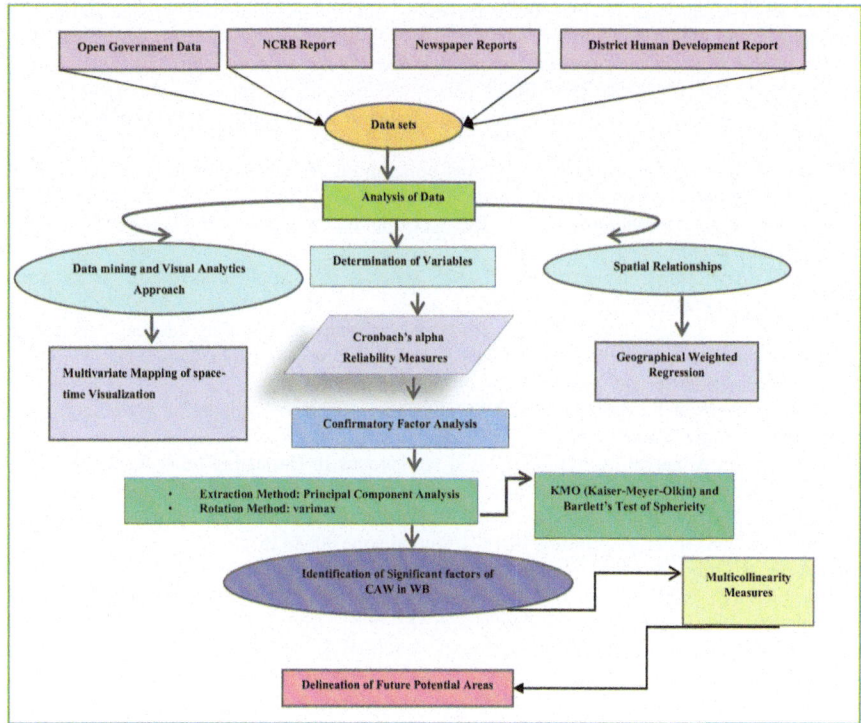

Fig. 1.4 Methodological overview

References

Anderson, J. C., & Gerbing, D. W. (1988). Structural equation modeling in practice: A review and recommended two-step approach. Psychological bulletin, 103(3), 411.

Anselin, L. (1998). Exploratory spatial data analysis in a geocomputational environment. In P. A. Longley, S. M. Brooks, R. McDonnelln, & B. Macmillan (Eds.), *Geocomputation: A primer* (pp. 77–94). Wiley.

Anselin, L. (2013). *Spatial econometrics: Methods and models* (Vol. 4). Springer Science & Business Media.

Anselin, L., Cohen, J., Cook, D., Gorr, W., & Tita, G. (2000). Spatial analyses of crime. *Criminal Justice, 4*(2), 213–262.

Anselin, L., Sridharan, S., & Gholston, S. (2007). Using exploratory spatial data analysis to leverage social indicator databases: The discovery of interesting patterns. *Social Indicators Research, 82*, 287–309.

Anselin, L., Meyer, W. D., Whalley, L. A., & Savoie, M. J. (2009). *Actionable cultural understanding for support to tactical operations (ACUSTO)): Toward a new methodological template for spatial decision support system*. US Army Corps of Engineers. ERDC/CERL TR-09-13.

Behrens, J. T. (1997). Principles and procedures of exploratory data analysis. *Psychological Methods, 2*(2), 131.

Bernasco, W., & Elffers, H. (2010). Statistical analysis of spatial crime data. In A. R. Piquero & D. Weisburd (Eds.), *Handbook of quantitative criminology* (pp. 699–724). Springer.

Biswas, P., & Chatterjee, N. D. (2021). Environmental vulnerability and women trafficking: Exploring the Bengal sundarban deltaic region of India. In *Practices in regional science and sustainable regional development: Experiences from the Global South* (pp. 279–296).

Biswas, P., & Das Chatterjee, N. (2024). Misery of the dark world: assessing risk of young women trafficking for commercial sexual exploitation in Darjeeling Himalayas using fuzzy TOPSIS. *SN Social Sciences, 4*(4), 82.

Braga, A. A. (2001). The effects of hot spots policing on crime. *The Annals of the American Academy of Political and Social Science, 578*(1), 104–125.

Brantingham, P. J., & Brantingham, P. L. (Eds.). (1981). *Environmental criminology* (pp. 27–54). Sage Publications.

Brown, T. A. (2006). Confirmatory factor analysis for applied research. New York, NY. Guilford Press.

Cerny, B. A., & Kaiser, H. F. (1977). A study of a measure of sampling adequacy for factor-analytic correlation matrices. *Multivariate Behavioral Research, 12*(1), 43–47.

Chainey, S., & Ratcliffe, J. (2005). *GIS and crime mapping*. Wiley.

Chainey, S., Tompson, L., & Uhlig, S. (2008). The utility of hotspot mapping for predicting spatial patterns of crime. *Security Journal, 21*, 4–28.

Chakrabarti, M., & Sarkar, A. (2007). Closed tea gardens in Darjeeling hills: A case study. In *Behind closed and abandoned tea gardens-status report of India* (p. 28).

Chen, Y. (2013). New approaches for calculating Moran's index of spatial autocorrelation. *PLoS One, 8*(7), e68336. https://doi.org/10.1371/journal.pone.0068336

Cliff, A. D., & Ord, J. K. (1981). *Spatial processes: Models & applications*. Pion.

Cohen, L. E., & Felson, M. (1979). Social change and crime rate trends: A routine activity approach. *American Sociological Review, 44*, 588–608.

Cohen, M. A., Rust, R. T., Steen, S., & Tidd, S. T. (2004). Willingness-to-pay for crime control programs. *Criminology, 42*(1), 89–110.

Craglia, M., Haining, R., & Wiles, P. (2000). A comparative evaluation of approaches to urban crime pattern analysis. *Urban Studies, 37*(4), 711–729.

da Silva, A. R. C., de Paula Júnior, I. C., da Silva, T. L. C., de Macêdo, J. A. F., & Silva, W. C. P. (2020). Prediction of crime location in a Brazilian city using regression techniques. In *2020 IEEE 32nd international conference on tools with artificial intelligence (ICTAI)* (pp. 331–336). IEEE.

Dou, Y., Luo, X., Dong, L., Wu, C., Liang, H., & Ren, J. (2016). An empirical study on transit-oriented low-carbon urban land use planning: Exploratory Spatial Data Analysis (ESDA) on Shanghai, China. *Habitat International, 53*, 379–389.

Eck, J. E., Chainey, S., Cameron, J. G., Leitner, M., & Wilson, R. E. (2005). *Mapping crime: Understanding hot spots*. In NIJ special report. https://www.ojp.gov/pdffiles1/nij/209393.pdf

Felson, M., & Poulsen, E. (2003). Simple indicators of crime by time of day. *International Journal of Forecasting, 19*(4), 595–601.

Fitterer, J., Nelson, T. A., & Nathoo, F. (2015). Predictive crime mapping. *Police Practice and Research, 16*(2), 121–135.

Fotheringham, A. S., Brunsdon, C., & Charlton, M. E. (2009). Geographically weighted regression. In *The Sage handbook of spatial analysis* (Vol. 1, pp. 243–254).

Gelman, A. (2004). Exploratory data analysis for complex models. *Journal of Computational and Graphical Statistics, 13*(4), 755–779.

Ghosh, B. (2014). Vulnerability, forced migration and trafficking in children and women: A field view from the plantation industry in West Bengal. *Econ Political Weekly, 49*(26), 58–65.

Gorman, D. M., Speer, P. W., Gruenewald, P. J., & Labouvie, E. W. (2001). Spatial dynamics of alcohol availability, neighborhood structure and violent crime. *Journal of Studies on Alcohol, 62*(5), 628–636.

Gorr, W., Olligschlaeger, A., & Thompson, Y. (2003). Short-term forecasting of crime. *International Journal of Forecasting, 19*(4), 579–594.

References

GOSCHIN, Z. (2019). Local factors explaining the incidence of criminal offences in Romania. A geographically weighted regression model. *Romanian Statistical Review, 2*.

Grubesic, T. H. (2006). On the application of fuzzy clustering for crime hot spot detection. *Journal of Quantitative Criminology, 22*, 77–105.

Guerry, A. M. (1833). *Essai sur la statistique morale de la France*. Paris, France.

Hagenauer, J., Helbich, M., & Leitner, M. (2011). Visualization of crime trajectories with self-organizing maps: a case study on evaluating the impact of hurricanes on spatio-temporal crime hotspots. In *Proceedings of the 25th conference of the international cartographic association, Paris*.

Harries, K. D. (1974). *The geography of crime and justice*. McGraw-Hill.

He, L., Páez, A., & Liu, D. (2017). Persistence of crime hot spots: an ordered probit analysis. *Geographical Analysis, 49*(1), 3–22.

Henry, L. M., & Bryan, B. A. (2000). Visualizing the Spatio-temporal patterns of motor vehicle theft in Adelaide. South Australia. In: National Centre for Social Applications of GIS (GISCA).

Hunt, E. D., Sumner, M., Scholten, T. J., & Frabutt, J. M. (2008). Using GIS to identify drug markets and reduce drug-related violence. In Y. F. Thomas, D. Richardson, & I. Cheung (Eds.), *Geography and drug addiction* (pp. 395–413). Springer.

Jiang, C., Liu, L., Qin, X., Zhou, S., & Liu, K. (2021). Discovering spatial-temporal indication of crime association (STICA). *ISPRS International Journal of Geo-Information, 10*(2), 67.

Johnson, S. D., & Bowers, K. J. (2004). The stability of space-time clusters of burglary. *British Journal of Criminology, 44*(1), 55–65.

Johnson, S. D., & Bowers, K. J. (2008). Stable and fluid hotspots of crime: Differentiation and identification. *Built Environment, 34*(1), 32–45.

Kabiraj, P. (2023). Crime in India: A spatio-temporal analysis. *GeoJournal, 88*(2), 1283–1304.

Komorowski, M., Marshall, D. C., Salciccioli, J. D., & Crutain, Y. (2016). Exploratory data analysis. In *Secondary analysis of electronic health records*. Springer. https://doi.org/10.1007/978-3-319-43742-2_15

Krivoruchko, K., Gotway, C. A., & Zhigimont, A. (2003, November). Statistical tools for regional data analysis using GIS. In *Proceedings of the 11th ACM international symposium on Advances in geographic information systems* (pp. 41–48).

Kumar, J., Mishra, S., & Tiwari, N. (2012). Identification of hotspots and safe zones of crime in Uttar Pradesh, India: Geo-spatial analysis approach. *International Journal of Remote Sensing Applications, 2*(1), 15–19.

LeBeau, J. L., & Leitner, M. (2011). Introduction: Progress in research on the geography of crime. *The Professional Geographer, 63*(2), 161–173.

Levine, N. (2006). The CrimeStat program: Characteristics, use and audience. *Geographical Analysis, 38*(1), 41–56. https://doi.org/10.1111/j.0016-7363.2005.00673

Liu, L., & Eck, J. (Eds.). (2008). *Artificial crime analysis systems: using computer simulations and geographic information systems: Using computer simulations and geographic information systems*. IGI Publishing.

Lu, B., Charlton, M., Harris, P., & Fotheringham, A. S. (2014). Geographically weighted regression with a non-Euclidean distance metric: A case study using hedonic house price data. *International Journal of Geographical Information Science, 28*(4), 660–681.

MacIndoe, H., & Oakley, D. (2023). Encouraging a spatial perspective in third sector studies: Exploratory spatial data analysis and spatial regression analysis. *Voluntas: International Journal of Voluntary and Nonprofit Organizations, 34*(1), 64–75.

Majumder, S., Roy, S., Bose, A., & Chowdhury, I. R. (2023). Multiscale GIS based-model to assess urban social vulnerability and associated risk: Evidence from 146 urban centers of Eastern India. *Sustainable Cities and Society, 96*, 104692.

Messner, S. F., & Anselin, L. (2004). Spatial analyses of homicide with areal data. In M. F. Goodchild & D. G. Janelle (Eds.), *Spatially integrated social science* (Vol. 12, pp. 127–144). Oxford University Press.

Messner, S. F., Anselin, L., Baller, R. D., Hawkins, D. F., Deane, G., & Tolnay, S. E. (1999). The spatial patterning of county homicide rates: An application of exploratory spatial data analysis. *Journal of Quantitative Criminology, 15*, 423–450.

Messner, S. F., Teske, R. H., Jr., Baller, R. D., & Thome, H. (2013). Structural covariates of violent crime rates in Germany: Exploratory spatial analyses of Kreise. *Justice Quarterly, 30*(6), 1015–1041.

Mohler, G. O., Short, M. B., & Brantingham, P. J. (2017). The concentration dynamics tradeoff in crime hot spotting. In D. Weisburd & J. Eck (Eds.), *Unraveling the crime-place connection* (Vol. 22, pp. 19–40). Taylor & Francis.

Molinari, N. (2017). Intensifying Insecurities: The impact of climate change on vulnerability to human trafficking in the Indian Sundarbans. *Anti-Trafficking Review, 8*.

Morenoff, J. D., Sampson, R. J., & Raudenbush, S. W. (2001). Neighborhood inequality, collective efficacy, and the spatial dynamics of urban violence. *Criminology, 39*(3), 517–558.

Mosteller, F., & Tukey, J. W. (1977). Data analysis and regression. A second course in statistics. *Addison-Wesley series in behavioral science: Quantitative methods*.

Murray, A. T., & Estivill-Castro, V. (1998). Cluster discovery techniques for exploratory spatial data analysis. *International Journal of Geographical Information Science, 12*(5), 431–443.

Murray, A. T., McGuffog, I., Western, J. S., & Mullins, P. (2001). Exploratory spatial data analysis techniques for examining urban crime: Implications for evaluating treatment. *British Journal of Criminology, 41*(2), 309–329.

Nunnally, J.C. (1994) Psychometric Theory 3E. Tata McGraw-Hill Education.

Ratcliffe, J. H. (2004a). Crime mapping and the training needs of law enforcement. *European Journal on Criminal Policy and Research, 10*(1), 65–83.

Ratcliffe, J. H. (2004b). The hotspot matrix: A framework for the spatio-temporal targeting of crime reduction. *Police Practice and Research, 5*(1), 5–23.

Ratcliffe, J. (2012). *Intelligence-led policing* (p. 288). Willan Publishing.

Rengert, G. F. (1997). Auto theft in central Philadelphia. In R. Homel (Ed.), *Policing for prevention: Reducing crime, public intoxication and injury* (Vol. 7, pp. 199–219). Criminal Justice Press.

Roy, S., & Chowdhury, I. R. (2023a). Geography of crime against women in West Bengal, India: Identifying spatio-temporal dynamics and hotspots. *GeoJournal, 88*(6), 5863–5895.

Roy, S., & Chowdhury, I. R. (2023b). Three decades of GIS application in spatial crime analysis: Present global status and emerging trends. *The Professional Geographer, 75*(6), 882–904.

Santos, R. B. (2016). *Crime analysis with crime mapping*. Sage.

Seltman, H. J. (2012). *Experimental design and analysis*. http://www.stat.cmu.edu/~hseltman/309/Book/Book.pdf. Last accessed 21 May 2023.

Seltman, H. J. (2018). Experimental design and analysis. *Book is on the World Wide Web*.

Shafeeq, A., Binu, V., & Binu, V. (2014). Spatial patterns of crimes in India using data mining techniques. *International Journal of Engineering and Innovative Technology, 3*(11), 291–295.

Shaw, C. R., & McKay, H. D. (1942). *Juvenile delinquency and urban areas*. University of Chicago Press.

Sherman, L. W., & Weisburd, D. (1995). General deterrent effects of police patrol in crime "hot spots": A randomized, controlled trial. *Justice Quarterly, 12*(4), 625–648.

Sherman, L. W., Gartin, P. R., & Buerger, M. E. (1989). Hot spots of predatory crime: Routine activities and the criminology of place. *Criminology, 27*(1), 27–56.

Suhr, D. D. (2006). Exploratory or confirmatory factor analysis? SUGI 31 Proceedings, 1–17. https://doi.org/10.1002/da.20406.

Tan, S. Y., & Haining, R. (2009). An urban study of crime and health using an exploratory spatial data analysis approach. In *Computational science and its applications–ICCSA 2009: International conference, proceedings, Part I* (Vol. 9, pp. 269–284). Springer.

Townsley, M. (2008). Visualising space time patterns in crime: the hotspot plot. *Crime Patterns and Analysis, 1*(1), 61–74.

Townsley, M., Homel, R., & Chaseling, J. (2000). Repeat burglary victimisation: Spatial and temporal patterns. *Australian & New Zealand Journal of Criminology, 33*(1), 37–63.

References

Tukey, J. W. (1977). *Exploratory data analysis* (Vol. 2, pp. 131–160). Addison-Wesley.

Wang, F. (2012). Why police and policing need GIS: An overview. *Annals of GIS, 18*(3), 159–171.

Weisburd, D., Bushway, S., Lum, C., & Yang, S. M. (2004). Trajectories of crime at places: A longitudinal study of street segments in the city of Seattle. *Criminology, 42*(2), 283–322.

Weisburd, D., Morris, N. A., & Groff, E. R. (2009). Hot spots of juvenile crime: A longitudinal study of arrest incidents at street segments in Seattle, Washington. *Journal of Quantitative Criminology, 25*(4), 443–467.

Weisburd, D., Telep, C. W., & Braga, A. A. (2010). *The importance of place in policing: Empirical evidence and policy recommendations* (pp. 1–68). Swedish National Council for Crime Prevention.

Wheeler, D., & Tiefelsdorf, M. (2005). Multicollinearity and correlation among local regression coefficients in geographically weighted regression. *Journal of Geographical Systems, 7*(2), 161–187.

Wu, X., & Grubesic, T. H. (2010). Identifying irregularly shaped crime hot-spots using a multiobjective evolutionary algorithm. *Journal of Geographical Systems, 12*(4), 409–433.

Zakaria, S., & Rahman, N. A. (2015). Analyzing the violent crime patterns in Peninsular Malaysia: Exploratory spatial data analysis (ESDA) approach. *Jurnal Teknologi (Sciences & Engineering), 72*(1), 131–136.

Chapter 2
Brief Elaboration of District-Wise Socioeconomic Settings in West Bengal

Introduction

To understand crime mechanisms, knowledge of the core situational environment is indispensable for practitioners to execute proper mitigating strategies to prevent crime in any given area. Exact space-specific factors, say specific physical and socioeconomic environmental settings, sometimes heighten unfavorable situations and eventually give rise to socioeconomic adversities in that area. Adverse physiographic settings, unfavorable infrastructural setups, and poor socioeconomic conditions sometimes foster poor social networks, the breakdown of social ties, the deepening of socioeconomic inequalities, and antisocial behavior in society, thus facilitating the upward trajectory of unsustainable regional development. Hence, to understand the prevalence of a criminogenic environment in a region and enact effective measures, a sound understanding of the physiographic settings as well as the associated socioeconomic backcloth is needed. In this context, the present chapter aims to provide a cogent understanding of the physiographic as well as socioeconomic and demographic aspects of the state of West Bengal to offer a glimpse of the situational environment that raises women's risk in society and makes them susceptible to crime.

Administrative Setup of West Bengal

West Bengal is India's 13th-largest state (covering an area of 88,752 square kilometers/34,267 square miles) with diversified physiographical settings. It stretches from the Darjeeling Himalayan hilly region in the north to the Bay of Bengal in the south. Part of the Bengal region of the Indian subcontinent, West Bengal shares international borders with Bangladesh to the east and Bhutan and Nepal to the north. It also

Table 2.1 Administrative divisions of West Bengal

Administrative division	Presidency division	Medinipur division	Burdwan division	Malda division	Jalpaiguri division
Name of districts	Howrah	Bankura	Birbhum	Malda	Alipurduar
	Kolkata	Jhargram	Paschim Bardhaman	Dakshin Dinajpur	Darjeeling
	Nadia	Paschim Medinipur	Purba Bardhaman	Uttar Dinajpur	Cooch Behar
	North-24 Parganas	Purulia	Hooghly	Murshidabad	Kalimpong
	South-24 Parganas	Purba Medinipur			Jalpaiguri

shares national borders with five Indian states: Bihar, Orissa, Jharkhand, Sikkim, and Assam. The state capital of West Bengal is Kolkata, which is the third-largest urban agglomeration and the seventh-largest city (by population) in India. West Bengal is divided into five administrative divisions: the Presidency Division, Medinipur Division, Burdwan Division, Malda Division, and Jalpaiguri Division. Each division comprises a group of districts and is headed by a "Divisional Commissioner." The districts under each division are shown in Table 2.1.

Demography and Socioeconomic Environmental Settings in West Bengal

The demographic structure of West Bengal is largely controlled by its physiographic settings, making this state vulnerable in many aspects. The porous international border with neighboring countries highly contributes to large-scale immigration and makes it the most densely populated state in India. In terms of demography, West Bengal is the fourth-most populous state in India, as per the 2011 census (with over 9.13 crore population). It accounts for 7.54% of India's total population. The percentage share of the male and female population in West Bengal (as per the 2011 census) is 51.28 and 48.72%, respectively.

As per the 2011 census, the sex ratio of West Bengal is 947 females per 1000 males, and the child sex ratio is 956 female children per 1000 male children. As of 2011, West Bengal is considered the second-most densely populated state in India after Bihar, with a population density of 1029 inhabitants per square kilometer. During 2011, the districts of Kolkata, North-24 Parganas, Howrah, and Hooghly had the highest population density, while Bankura, Purulia, Birbhum, Dakshin Dinajpur, Darjeeling, and Jalpaiguri had the lowest. Specifically, the border districts of West Bengal are densely populated due to huge immigration from neighboring country Bangladesh. The decennial growth rate of the population in the 2001–2011 decade was 13.84%, a slight decrease of 4.01% compared to the previous decade

(17.84% in 1991–2001) and lower than the national rate (17.64%). The social behavioral aspects are also influenced by the demographic patterns and the socioeconomic settings of the state. The huge population pressure is reflected in the social structure, education system, economy, social behavior, social network, neighborhood behavioral pattern, and so on.

As per the 2011 census, the literacy rate of West Bengal is 76.26%, with a male literacy rate of 81.69% and a female literacy rate of 70.54%. The rural-urban population share in West Bengal, as per the 2011 census, is 68.13 and 31.87%, respectively. Kolkata is a completely urbanized district in West Bengal. Meanwhile, North-24 Parganas and Howrah have 57.27 and 63.38% urban populations, respectively (as per the census 2011). The percentage share of Scheduled Caste (SC) and Scheduled Tribe (ST) populations in West Bengal is 27.49 and 7.81% in rural areas and 15.01 and 1.52% in urban areas, respectively. The backward districts in terms of SC and ST populations in West Bengal are, as per the 2011 census, North- and South-24 Parganas, Bardhaman, Nadia, Jalpaiguri, and Cooch Behar for the SC population category and Paschim Medinipur, Jalpaiguri, and Purulia for the ST population category. The proportion of people living below the poverty level (BPL) based on a 2013 record was 19.98%, lower than the previous decade. However, no matter how much it reflects a positive situation, poverty itself is still a very burning issue in West Bengal. Regional disparities and economic imbalances are very much predominant. As of 2015–116, data from the Ministry of Statistics and Programme Implementation indicated that West Bengal's GSDP (gross state domestic product) share in India's GDP (gross domestic product) was 5.5%, showing a decrease in rate as compared to 2011–2012 (6.0%).

The state's economy largely depends on agriculture. Many people from rural Bengal depend on agriculture, and most of the farmers work as marginal workers. Paddy is the principal crop, which is mainly produced in the middle and southern Gangetic plain regions of West Bengal. North Bengal is well known for its high-quality tea production. Industries are localized mainly in Kolkata and its suburban region, specifically along the river side of Hooghly (Barrackpore, Uttar Para, Hind Motor, Serampore, Hooghly, Bandle, etc.), the mineral-rich western plateau highland region (Birbhum, Durgapur-Asansol colliery belt), and the Haldia port region.

Thus, it can be noted that economic development is mainly concentrated in some specific pockets of West Bengal and is not evenly distributed across the state, so regional imbalances are very much apparent. People from backward areas migrate to these comparatively developed regions for economic opportunities, damaging the social structure in backward areas and eventually breaking social ties. This uneven distribution of economy results in large-scale disguised unemployment in Bengal. These social insecurities further exacerbate social vulnerability and make society highly disorganized. In such disorganized environments in West Bengal, criminogenic activities have emerged dramatically in the recent past, and crime against women is one of them.

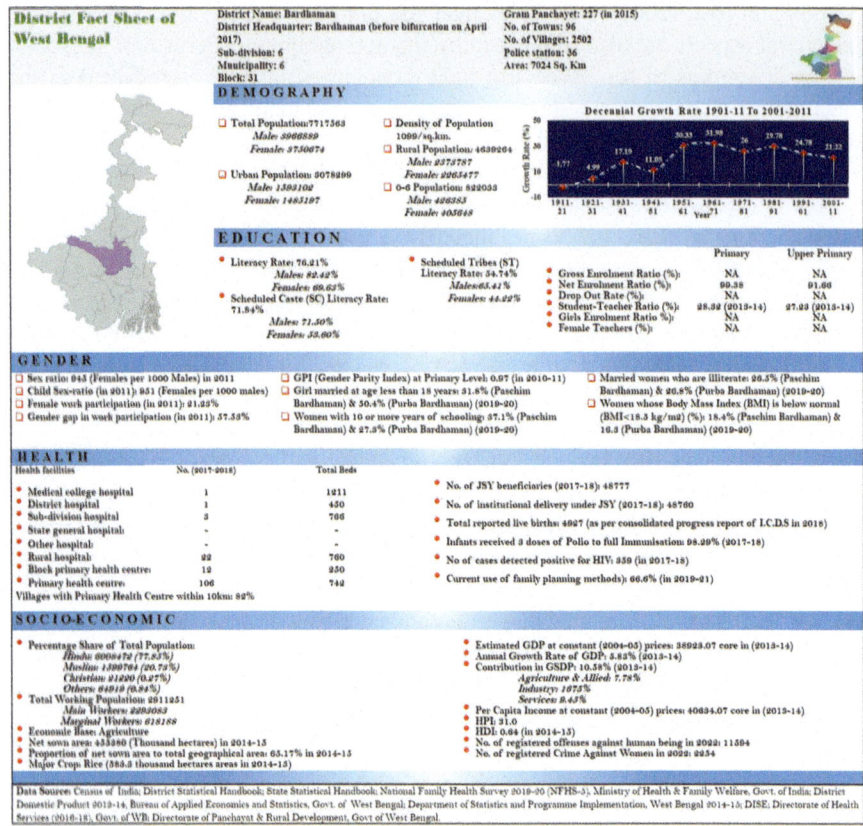

Fig. 2.1 District fact sheet of Bardhaman, West Bengal, India

The hostile physical environmental settings as well as the poor socioeconomic setup create such a situational environment that exacerbates women's susceptibilities in Bengal in many ways. District-wise dashboards (Figs. 2.1, 2.2, 2.3, 2.4, 2.5, and 2.6) are provided below to provide a brief overview of the socioeconomic environmental settings of some vulnerable districts in West Bengal.

Demography and Socioeconomic Environmental Settings in West Bengal

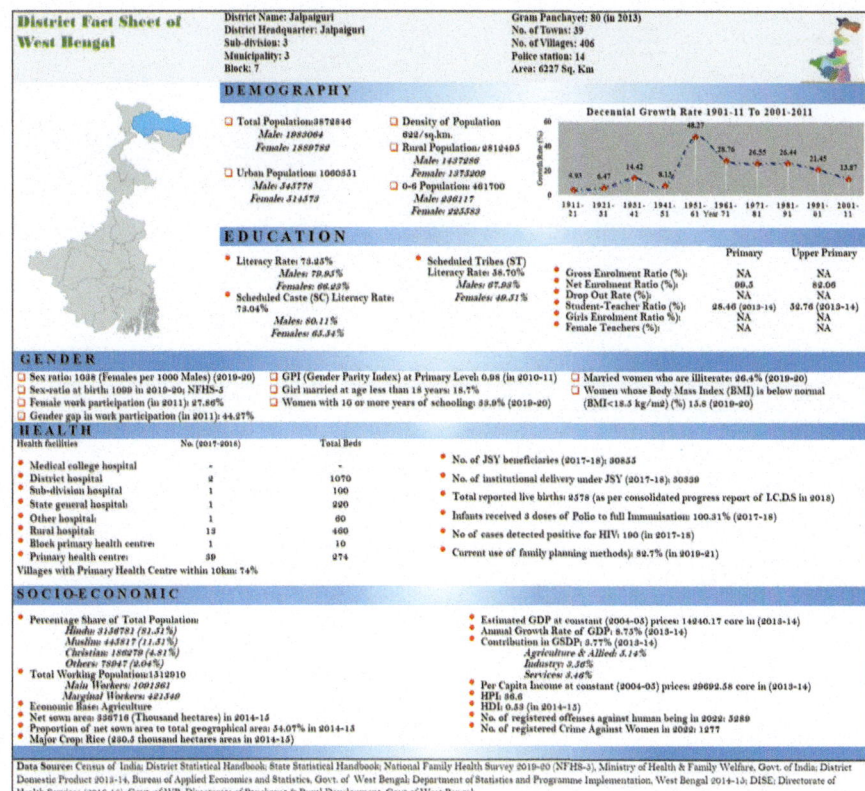

Fig. 2.2 District fact sheet of Jalpaiguri, West Bengal, India

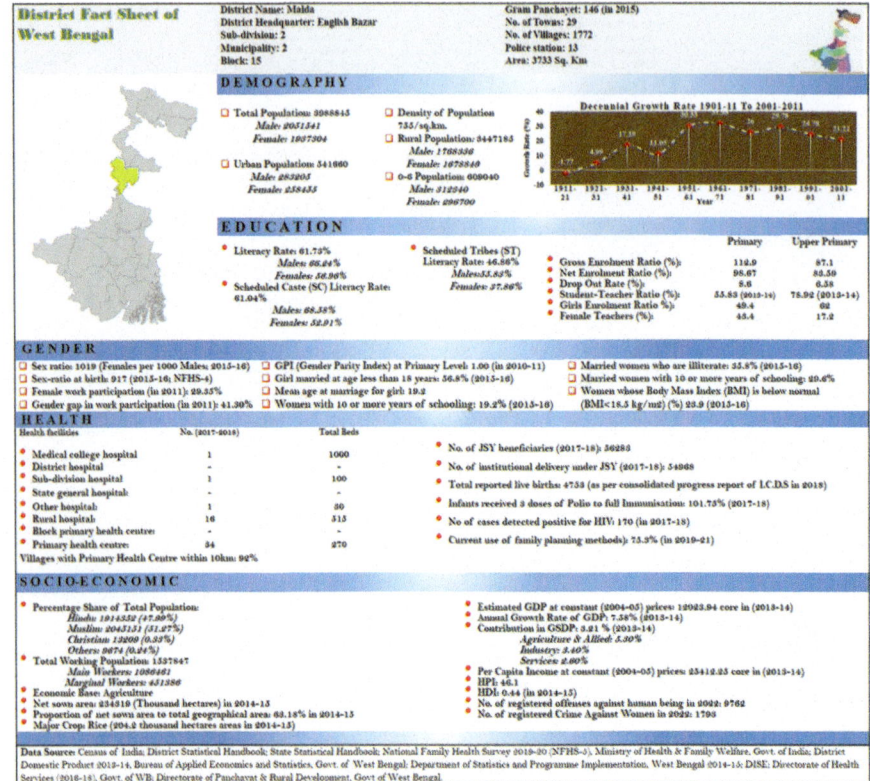

Fig. 2.3 District fact sheet of Malda, West Bengal, India

Demography and Socioeconomic Environmental Settings in West Bengal

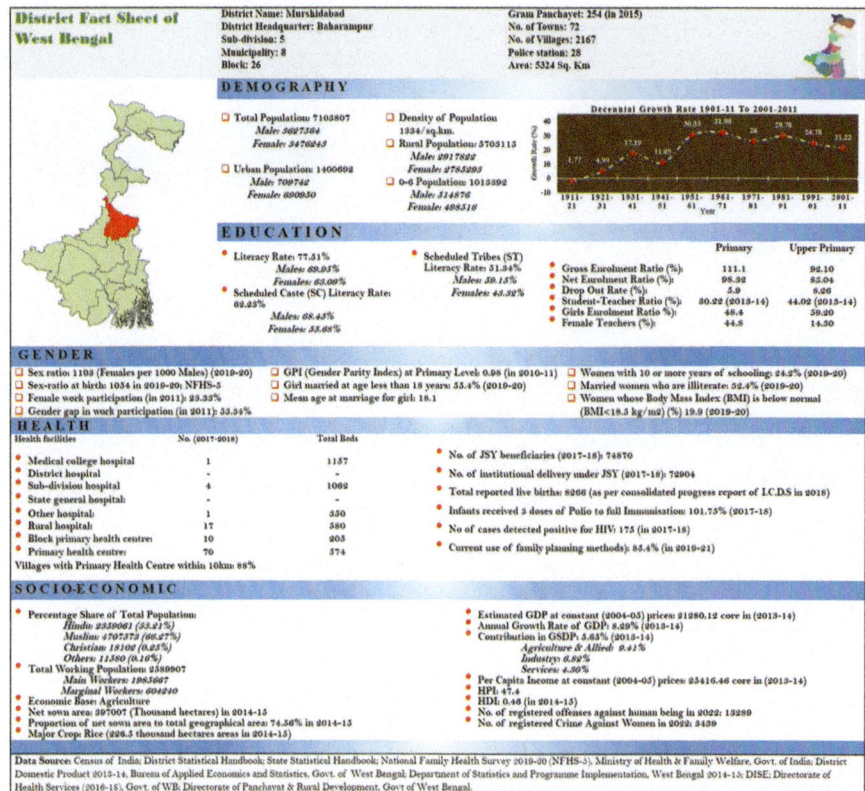

Fig. 2.4 District fact sheet of Murshidabad, West Bengal, India

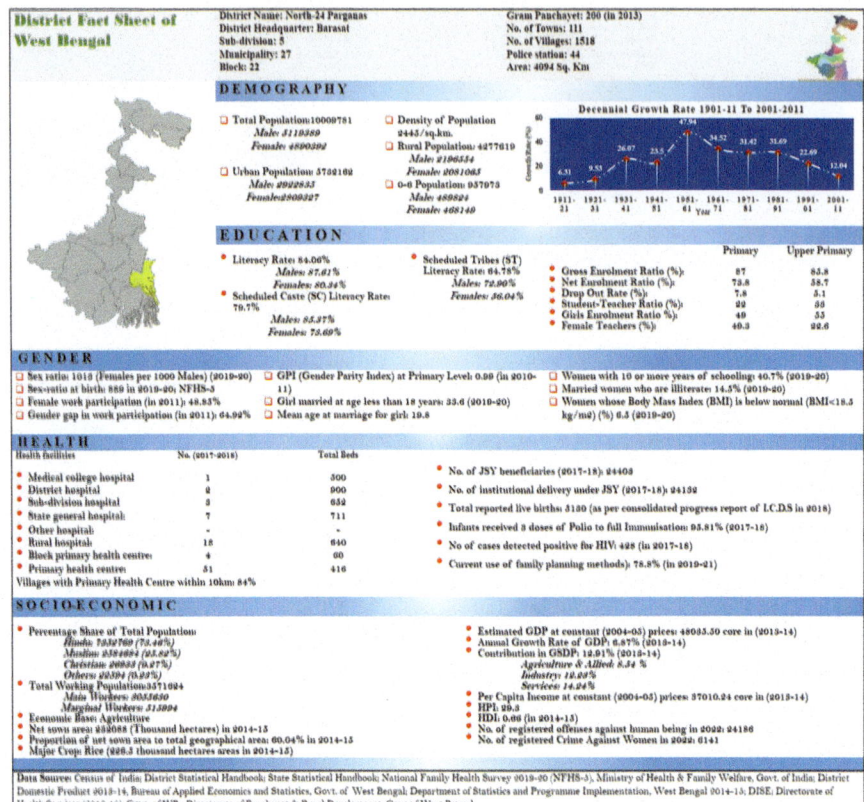

Fig. 2.5 District fact sheet of North-24 Parganas, West Bengal, India

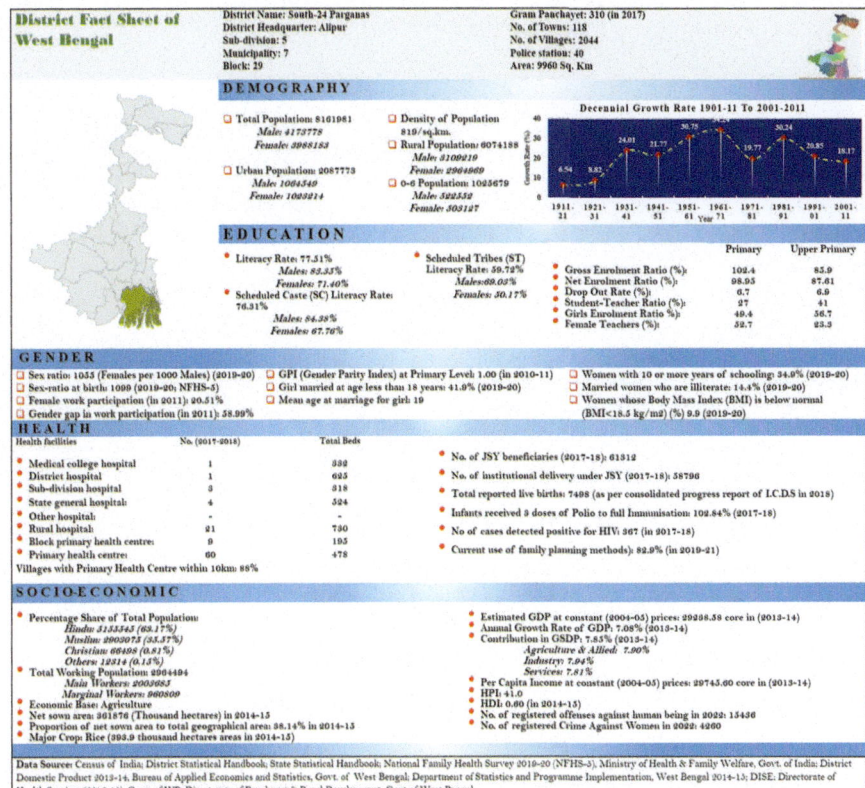

Fig. 2.6 District fact sheet of South-24 Parganas, West Bengal, India

Chapter 3
Geovisualization and Prediction of Crime Against Women in West Bengal Using Statistical Modeling

Exploratory Spatial Analysis of Crime Data

To gain insights into crime patterns from multiple perspectives (viz., crime composition, spatial extent, temporal evolvement, etc.) and to support place-based policing and planning efforts, the adoption of geovisualization techniques and statistical observation is noteworthy. The integration of statistical methods with Geographic Information System (GIS) tools for mapping and spatially modeling crime data is indispensable among crime analysts. Implementing statistics within GIS provides a powerful set of interactive, analytical tools uniquely suited to understanding the spatial association of data, comprehending crime patterns, and endorsing evidence-based policy implementation. Therefore, using district-level crime data, the present chapter aims to provide a comprehensive understanding of the spatiotemporal evolvement of all forms of crimes committed against women in West Bengal, India, to suit the goals of regional data analysis. It also seeks to ascertain future crime trends to facilitate the proper implementation of crime prevention measures in crime-ridden zones.

This chapter is organized as follows: section "Method" briefly elaborates on the methodology and techniques for data processing. Section "Periodic insights into data: Multivariate mapping and space-time attribute visualizations" provides detailed periodic insights into data for multivariate mapping and space-time attribute visualization and the analysis of crime patterns. Section "Geospatial extent of crime concentration: Some case studies based on newspaper reports" statistically models entire crime patterns to reveal future trends for different crime forms. Finally, section "Results of spatial autocorrelation (Moran's I)" presents a robust discussion and offers future directions.

Method

The sources of district-level crime data, techniques for data arrangement and processing, and data normalization procedures are all explained in Chap. 1. In this section, a brief overview of the adopted methodology and data-processing techniques have been discussed to gain periodic insights into data and analyze crime patterns. After completing data normalization, thematic mapping, parallel coordinate plots (PCPs), and heatmap have been drawn simultaneously to more conveniently geovisualize district-wise spatial variation of different crime categories (considering four major crime categories from 2010 to 2022 for intense study, viz., cruelties by husband or his relatives (Sec. 498-A IPC), dowry deaths (Sec. 304 B IPC), rape (Sec. 376 IPC), and kidnapping and abduction (Secs. 363–373 IPC), which have already been mentioned in Chap. 1) across varying temporal ranges and identify the districts with similar temporal trends.

Gross crime data during the 12-year (2010–2022) time span has also been considered to gain periodic insights into crime patterns and women susceptibilities across West Bengal. In PCP, each row (district, census block, or neighborhood) in the data table is plotted as a line or profile, and each attribute (year-wise-registered crime data) of a row is plotted as a point on that line or profile. Afterward, *Moran's I* analysis is performed to highlight the spatial clustering patterns of different crime forms across West Bengal, providing statistical robustness to detect significant high- and low-value clustering areas of crime concentration.

After gaining periodic insights into different crime categories perpetrated against women in West Bengal from multivariate perspectives and understanding significant clustering patterns, crime forecasting with improved predictive accuracy has been conducted to support policing strategies. ARIMA (autoregressive integrated moving average) analysis based on rolling forecasting origin is performed for the next 5 years to forecast future crime trends in West Bengal. The detailed architecture and algorithm of the ARIMA model are later described. Finally, a robust discussion has been conducted in connection with future trends on the spatial extension of crime patterns to draw the attention of practitioners and policymakers to implement the best possible crime prevention measures in the most crime-ridden districts.

Periodic Insights into Data: Multivariate Mapping and Space-Time Attribute Visualizations

Even though the occurrence of crime may be random or clustered and very unpredictable, crime mapping and analysis, an important tool for exploratory spatial data analysis (ESDA), is very much accepted among practitioners for providing deep insights into given data to discern the spatial expansion and temporal evolution of crimes. ESDA is an important step toward geovisualizing crime patterns, clusters, and outliers within datasets (Baller et al., 2001). It provides a flexible way to observe data, generate hypotheses, and detect unexpected spatial patterns (Eck et al., 2005).

Periodic Insights into Data: Multivariate Mapping and Space-Time Attribute... 35

Using GIS proficiencies as an ESDA tool, this study analyzes the evolvement of crime patterns across space and time through multivariate mapping and space-time attribute visualization, including the map-matrix, heatmap, and PCP. The rows in the heatmap specify the districts (or neighborhood or census blocks), and the columns denote the time period from 2010 to 2022. Each column in the heatmap corresponds to a map in the map-matrix. More specifically, both the heatmap and the map-matrix reflect the same data in two distinct ways: the former reveals the temporal evolution of crime, while the latter simultaneously exhibits spatial and temporal extent. The PCP visualizes clusters of time series, which is ideal for determining districts with parallel temporal crime trends by comparing profiles and understanding their interrelation.

The map-matrix (Part A of Fig. 3.1) shows the temporal evolution of spatial extension patterns of cruelties by husband or his relatives (498-A IPC) in West

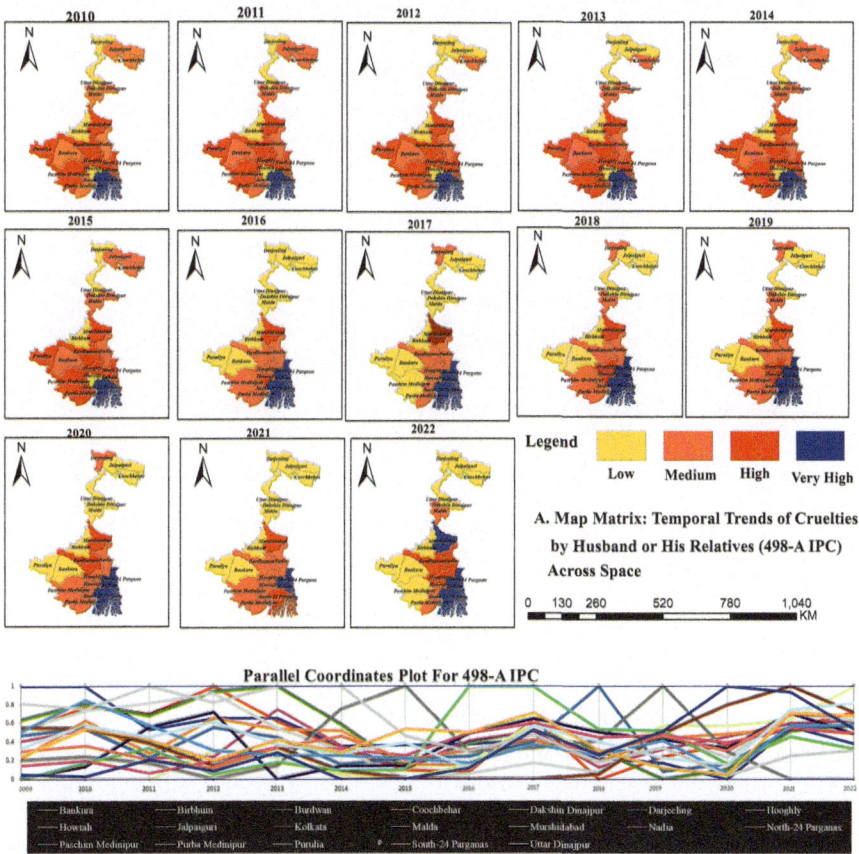

Fig. 3.1 Evolvement patterns of cruelties by husband or his relatives (498-A IPC) across space and time in West Bengal. The visualization comprises (**a**) a map-matrix, (**b**) a parallel coordinate plot, and (**c**) a heatmap. (Source: data.gov.in)

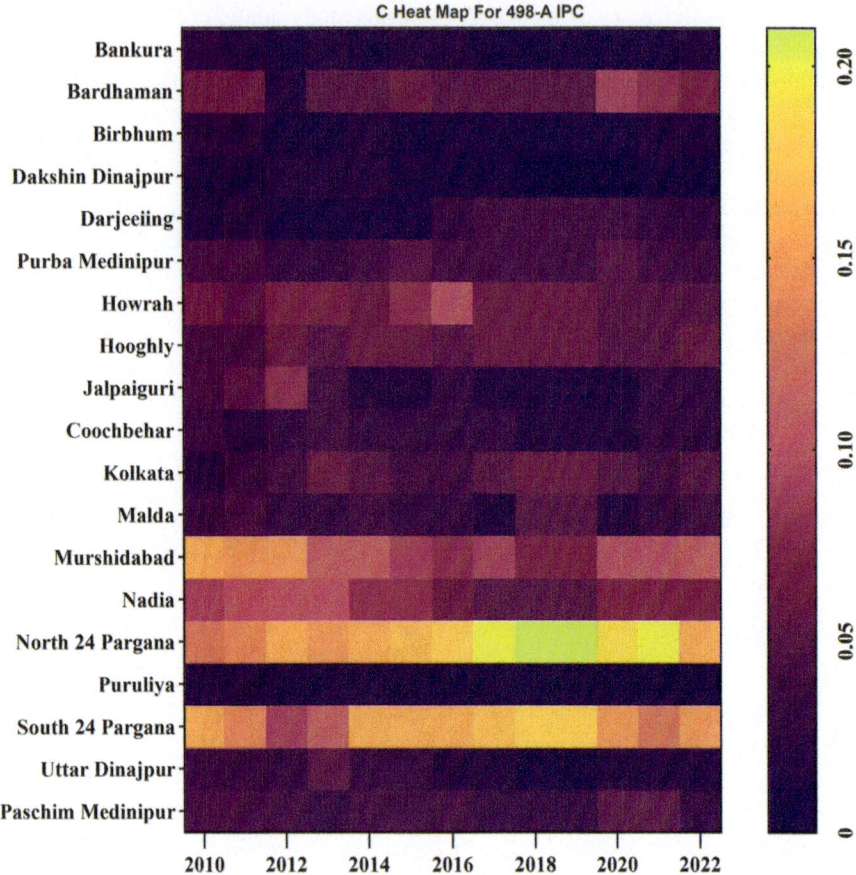

Fig. 3.1 (continued)

Bengal, from which it is revealed that the districts North- and South-24 Parganas, Paschim Medinipur, Murshidabad, Bardhaman, Malda, Hooghly, and Howrah are the most vulnerable in terms of victimization of women in domestic environments. In the districts of Bankura, Jalpaiguri, Cooch Behar, and Dakshin Dinajpur, a moderate pattern of domestic violence victimization was observed from 2010 to 2015; afterward, these districts experienced comparatively fewer reported incidents of domestic violence up to 2022. From 2010 to 2016, the district of Darjeeling reported very few occurrences of domestic violence victimization among women, but afterward, it held a moderate position, with a slightly rising trend till 2020. From 2021 to 2022, Darjeeling experienced fewer reported incidents of cruelties in domestic environments. From 2010 to 2016, a surging trend of high domestic violence victimization was noticed in the district of Purulia, but after 2016, a declining trend followed in this district. It is astonishing to see that the state capital, Kolkata, underreported incidents of domestic violence from 2010 to 2016 but afterward experienced a gradual rising trend till 2022. The map-matrix also shows that the districts of Purba Medinipur and Nadia held a moderate position and Uttar Dinajpur held a low position with regard to recorded incidents of domestic violence victimization among women during these 12 years.

From the PCP (Part B of Fig. 3.1), a clearer picture emerges regarding the clustering of districts concerning the temporal evolvement of cruelties in domestic environments in West Bengal. The PCP of domestic violence (498-A IPC) indicates that during the 12 years from 2010 to 2022, the districts of Murshidabad, North- and South-24 Parganas, Purba and Paschim Medinipur, Howrah, Bardhaman, and Hooghly formed a cluster with a surging trend in domestic violence victimization among ever-married women. In contrast to this, the districts of Darjeeling, Cooch Behar, Jalpaiguri, North and South Dinajpur, Birbhum, Bankura, and Purulia formed a cluster with a gradually declining trend, showing higher domestic violence victimization in earlier times compared to recent years.

Likewise, the heatmap (Part C of Fig. 3.1) geovisualizes that from 2010, the districts of North- and South-24 Parganas, Murshidabad, Malda, Bardhaman, Purba and Paschim Medinipur, Howrah, Nadia, and Hooghly have been in a most vulnerable position in terms of women's safety in domestic environments, constantly witnessing alarming incidences of cruelties during this time. Specifically, the district of South-24 Parganas witnessed a very high occurrence of domestic violence victimization among ever-married women throughout this time span. In the case of North-24 Parganas, it can be noted that since 2010, it has witnessed high to very high occurrences of domestic violence, with a very high pattern observed specifically after 2016. Surprisingly, however, the districts of Purulia, Bankura, Birbhum, North and South Dinajpur, Jalpaiguri, and Cooch Behar have, from 2016 onward, witnessed fewer occurrences of cruelties in domestic environments. This might be a reflection of lower reporting rates in these districts due to the improper execution of strong legislative measures, an ignorant attitude among women, unawareness about legal aids, and a general fear of the police.

Periodic insights into dowry deaths (304-B IPC) across space (districts) and time (year) in West Bengal have been revealed in Fig. 3.2. The map-matrix (Part A of Fig. 3.2) shows that from 2010 to 2022, the districts of Nadia, North- and South-24 Parganas, Murshidabad, Bardhaman, and Hooghly were most susceptible to occurrences of dowry deaths in West Bengal. In the district of Hooghly, initial records show moderate occurrences of dowry deaths, but from 2014 onward, the severity of occurrences gradually increased (except from 2018 to 2019, when occurrences were moderate). Official records also indicate that in prior times, the district of Cooch Behar initially held a moderate status concerning reported incidences of dowry deaths, but vulnerability gradually increased after 2020.

It has been shown from the map-matrix that from 2010 to 2015, the district of Purba Medinipur was in a moderately vulnerable position when it comes to dowry death cases. However, from 2016, the severity declined dramatically, though after 2019, Purba Medinipur became highly vulnerable again. The map-matrix also visualizes that in the districts of Jalpaiguri, Uttar Dinajpur, Malda, Bankura, Paschim

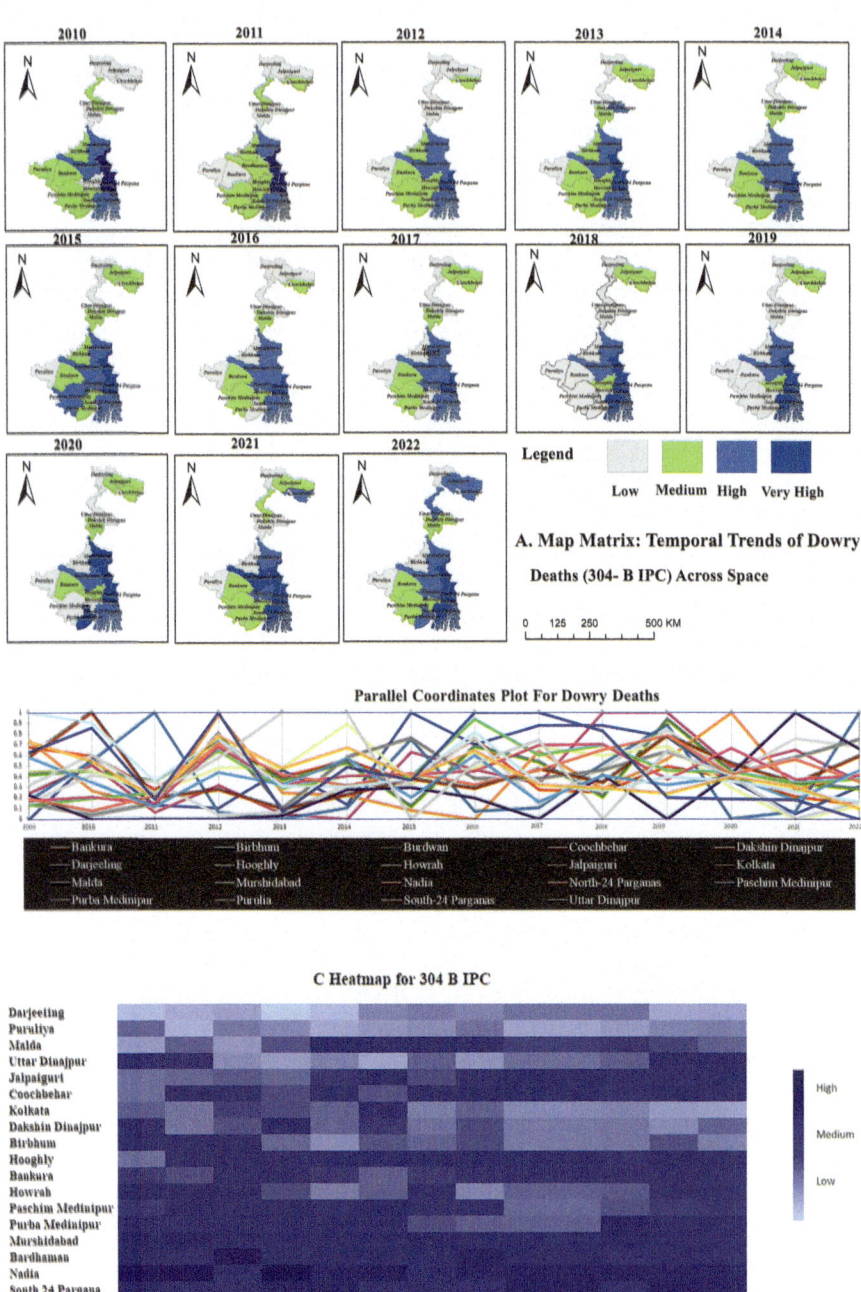

Fig. 3.2 Evolvement patterns of dowry deaths (304-B IPC) across space and time in West Bengal. The visualization comprises (**a**) a map-matrix, (**b**) a parallel coordinate plot, and (**c**) a heatmap. (Source: data.gov.in)

Medinipur, and Howrah, the severity of recorded incidences of dowry deaths was merely moderate during these 12 years.

The PCP (Part B of Fig. 3.2) exhibits clusters of districts with similar temporal trends of dowry deaths in West Bengal. It shows that from 2010 onward, a thoroughly surging trend of dowry deaths has been observed in the districts of North- and South-24 Parganas, Murshidabad, Nadia, Bardhaman, Hooghly, and Jalpaiguri in West Bengal. A steady rising and falling trend of dowry deaths has been observed since 2010 in the districts of Hooghly, Cooch Behar, Paschim and Purba Medinipur, and Uttar Dinajpur. From 2011 onward, an average temporal pattern of dowry death cases has been observed in the district of Cooch Behar. But after 2020, a steadily rising trend with a high intensity of registered dowry death cases followed in this district. From 2010 to 2015, a moderate trend was noticed in the district of Purba Medinipur, but after 2015, a gradual decreasing trend followed. Yet again, from 2020 onward, a surging trend of dowry deaths has been observed in this district. A decreasing trend with a moderate-to-low pattern of dowry deaths has been noted in the Uttar Dinajpur district from 2010 to 2020, though from 2021 onward, a gradual rising trend followed. In contrast to these, the PCP of dowry deaths also visualizes that an average temporal pattern with a gradually diminishing trend has been observed in the districts of Purulia, Birbhum, Dakshin Dinajpur, Darjeeling, and Kolkata during this 12-year time span. Therefore, a compact understanding of the correlation between the districts concerning the temporal evolution patterns of dowry deaths can easily be grasped through the PCP.

Composition patterns of dowry deaths across space and time are evidently reflected through the heatmap, too (Part C of Fig. 3.2). Based on the heatmap of dowry deaths, it is evident that from 2010 to 2022, the districts of North- and South-24 Parganas, Murshidabad, Nadia, Malda, Purba Medinipur, Bardhaman, Hooghly, Cooch Behar, and Jalpaiguri were at a high-risk position, with high to very high occurrences of dowry deaths. From 2010 to 2014, the district South-24 Parganas has been experienced very high occurrences of dowry deaths. However, from 2015 onward intensity was somewhat decreased, yet followed a high to moderately high trend. In the districts of Kolkata, Birbhum, and Dakshin Dinajpur, a gradually declining trend has been noticed, with more dowry death cases registered in earlier times compared to later years. And in the districts of Darjeeling and Purulia, an unceasingly parallel trend with a low-to-moderate intensity of dowry deaths has been observed through this heatmap.

District-wise spatiotemporal evolution patterns for the crime of rape (376 IPC) in West Bengal have been revealed in Fig. 3.3. The map-matrix (Part A of Fig. 3.3) shows that from 2010 onward, the districts of Malda, Murshidabad, North- and South-24 Parganas, Nadia, Bardhaman, Jalpaiguri, and Cooch Behar have been at high risk, witnessing high to very high reported incidences of rape against women. In 2016, a slightly decreasing trend of rape victimization in South-24 Parganas was noticed, but since 2017, this district again witnessed high to very high occurrences of rape against women. A parallel rising and falling trend of rape (with moderate-to-high occurrences of rape) against women has been noted in the districts of Bardhaman and Uttar Dinajpur during these 12 years. Likewise, in Hooghly and

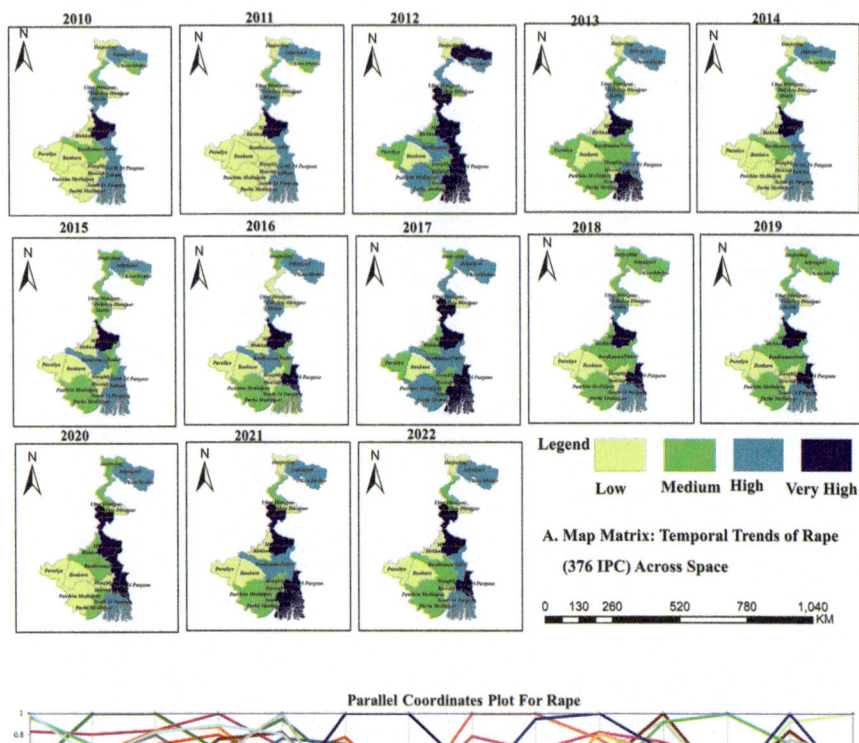
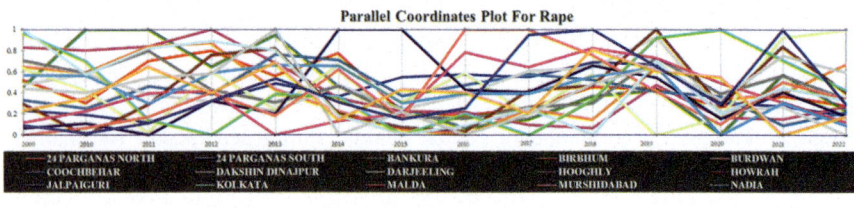
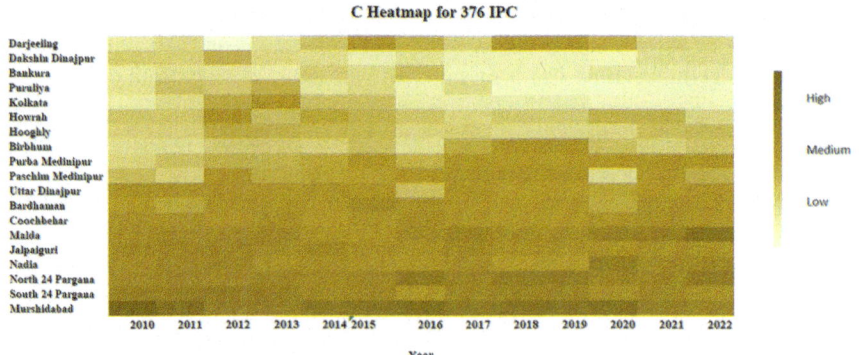

Fig. 3.3 Evolvement patterns of rape (376 IPC) across space and time in West Bengal. The visualization comprises (**a**) a map-matrix, (**b**) a parallel coordinate plot, and (**c**) a heatmap. (Source: data.gov.in)

Howrah, a parallel trend with low-to-moderate patterns was observed during this period. Official records revealed that in the earlier period, the Darjeeling district experienced a lower risk of rape against women, but since 2015, a rising pattern has been noted. In the Paschim Medinipur district, a moderate pattern of rape victimization has been observed since 2010, but the intensity of occurrences increased dramatically in 2012 and 2017. The map-matrix also reflects that since 2010, the districts Dakshin Dinajpur, Purulia, Bankura, and Kolkata have somewhat been less at risk with regard to rape against women.

The PCP (Part B of Fig. 3.3) reveals clusters of districts with similar spatiotemporal distributional patterns of rape in West Bengal. The districts of North- and South-24 Parganas, Murshidabad, Malda, Bardhaman, Cooch Behar, and Jalpaiguri form a cluster with a steadily rising trend since 2010. In Nadia, a parallel upward trend was observed from 2010 to 2014, which declined to a certain extent after that. But again, a surging trend with high to very high patterns has been observed in the later periods. Another clustering pattern with a steadily rising and dwindling trend of rape victimization has been noticed in the districts of Uttar Dinajpur, Hooghly, Howrah, Paschim and Purba Medinipur, and Darjeeling during the 12-year period. From 2010 to 2022, a low-to-moderate trend has been observed in the districts of Purba and Paschim Medinipur (except in 2012 and 2017, where an upward trend has been noted), with more rape cases registered in recent times. Another form of a cluster with fewer occurrences of rape has evolved from the PCP, which specifies that since 2010, Kolkata, Purulia, and Bankura have been experiencing comparatively fewer occurrences of rape against women in West Bengal.

The heatmap (Part C of Fig. 3.3) also exhibits a spatiotemporal evolvement pattern of rape in West Bengal, which also specifies the most crime-ridden districts when it comes to officially recorded incidences of rape against women in West Bengal: Murshidabad, North- and South-24 Parganas, Nadia, Jalpaiguri, Bardhaman, Paschim Medinipur, and Uttar Dinajpur, which have constantly witnessed a steady upheaving trend (high to very high) since 2010. In the Malda district, a moderate pattern with a slightly diminishing trend of rape victimization was observed from 2014 to 2015, but since 2016, vulnerability has increased again in this district. In North-24 Parganas, from 2013 to 2019, a steady rising and falling pattern of rape victimization was observed, but since 2020, severity has increased again in this district. Meanwhile, since 2010, the districts of Bankura, Purulia, Birbhum, and Hooghly have been in relatively low-risk positions, as reflected in the same heatmap.

Likewise, differential spatiotemporal evolution patterns of various forms of crimes in West Bengal from multiple perspectives might have emerged from the official data. For instance, from 2010 to 2022, occurrences of criminal activities such as "kidnapping and abduction" and "trafficking in young women and children" have been very high in some specific pockets of West Bengal, specifically in the border districts of Murshidabad, Malda, Nadia, North- and South-24 Parganas, Uttar Dinajpur, Dakshin Dinajpur, Cooch-Behar, and Darjeeling (to address the study objectives, not all crime visualization maps are shown here). So, from the periodic insights of official data, it is observed that crime committed against women in West Bengal is very space specific. Over the last decade, differential clusters of crime concentration have been formed for different types of crimes throughout West Bengal. However, the overall concentration of crime against women is highest in certain areas, specifically in the border districts of West Bengal.

Geospatial Extent of Crime Concentration: Some Case Studies Based on Newspaper Reports

Apart from glimpses of periodic insights into available official data, specific spatial clustering patterns of crime committed against women in West Bengal have evolved from analyzing some well-circulated newspaper archives. As the daily newspapers or e-news are full of news about incidences of dowry deaths, cruelty by husband and his relatives, the gruesome incidence of acid attacks, rape and other forms of sexual abuse, women trafficking and many other forms of gender-based atrocities against women, therefore some renowned newspaper records have been considered to overreach the geospatial extent of differential crimes committed against women in West Bengal. For instance, a clear portrait of the spatial extent of women trafficking in North and South-24 Parganas districts has evolved from the kernel density mapping (Fig. 3.4) based on published newspaper records on incidences of trafficking in some well-circulated dailies (viz., The Hindu, Times of India, The Telegraph, Anandabazar Patrika, Ei-Samay) (Appendix 1) from 2015 to 2018. This figure reveals that the inhumane practice of trafficking in young women and minor girls in the remote areas of North- and South-24 Parganas has been on the rise. The remote riverine poverty-stricken villages of Sundarban delta viz. the villages in Basanti, Gosaba, Canning I and II, Mathurapur I and II, Kakdwip, Joynagar I and II, Patharpatima, Kultali, Namkhana, and Sagar blocks of South-24 Parganas and Sandeshkhali I and II, Hasnabad, Hingalganj, Haroa, Minakhan blocks of North-24 Parganas district have emerged as the "major source and transit hub" of women trafficking. Apart from these poverty-stricken rural areas, other backward areas of these two districts viz., Mograhat, Enayet Nagar, Jibontala, Mandir Bazar, Diamond harbour, Malikpur, Taldi, and Sonarpur of South-24 Parganas and Bongaon, Ashok nagar, Habra, Bagda, Gaighata, Basirhat of North-24 Parganas serve as major hotbed of women trafficking in recent days. The physiographic disadvantageous locational settings of these riverine Sundarbans blocks, with an intricate network of creeks, isolate these regions from the rest of the civilization. In addition, frequent threats of climatic hazards (such as cyclones, tidal surges, sea-level rise, and floods in Sundarbans) create a wide swath between these riverine blocks and the mainland, resulting in enormous socioeconomic hardships, acute poverty, unemployment, lack of basic amenities, homelessness, starvation, and forcing people to migrate as "climate refugees" and making these regions highly susceptible. Traffickers take ultimate advantage of these worse situations and allure poor women and minor girls with false promises of better job prospects in metros, and eventually traffic them. The summarization of some newspaper reports in Appendix 1 concerning

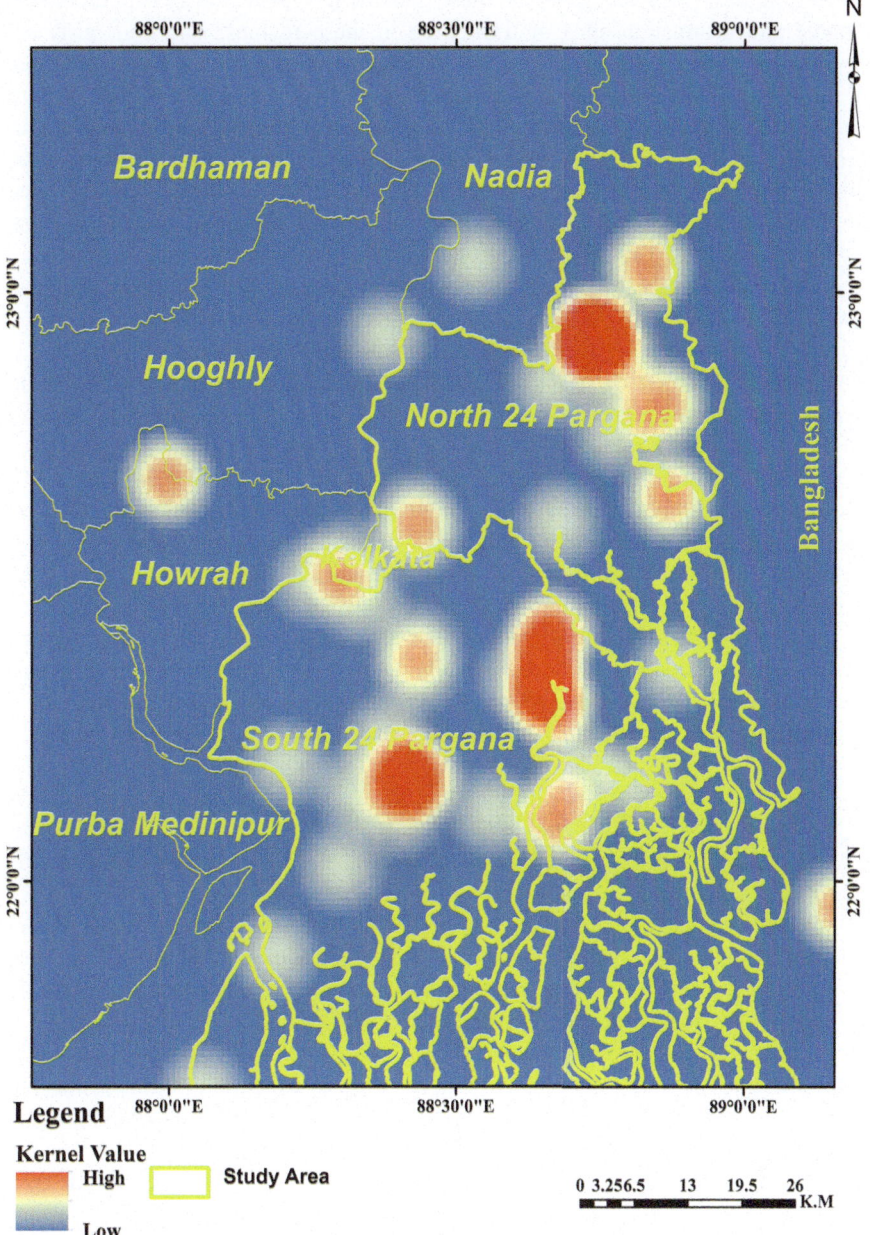

Fig. 3.4 Spatial extent of trafficking hotspots in North- and South-24 Parganas districts. (Source: Newspaper reports (The Hindu, Times of India, The Telegraph, Anandabazar Patrika, Ei-Samay))

trafficking in women and minors in these two districts meticulously explicit the key source and destination sites of trafficking in these regions. The poor girls and women from these remote villages are trafficked to metro cities in Delhi, Mumbai, Pune, Ahmadabad, Bangalore, Rajasthan, Bhopal, Bihar, etc., especially in red-light areas, and are forced to do filthy jobs. Some are sent abroad. The porous international border with the neighboring country Bangladesh facilitates large-scale illegal immigration, which eventually enables traffickers to execute such organized crime in a very convenient way. Another significant spatial clustering pattern of crime against women in West Bengal that has evolved by analyzing newspaper archives is the gruesome incidences of acid attacks against women, that scar many women's lives and condemn them to a fate worse than death. The kernel density mapping of Fig. 3.5 based on newspaper archives on incidences of acid attacks published in some well-circulated dailies (The Times of India, The Telegraph, The Hindu, Anandabazar Patrika) and Kolkata-based NGO ASFI's (Acid Survivors Foundation of India) records (Appendix 2) from 2010 to 2017 exhibits that South Bengal is comparatively more susceptible for such brutality against women than North Bengal. Canning, Baruipur of South-24 Parganas, Bongaon, Habra, Ashok nagar, Bagda, Gaighata of North-24-Parganas, Purbasthali of Bardhaman, Bagnan, Uluberia subdivision of Howrah, Ghatal, Daspur of Paschim Medinipur, even Kolkata, Purba Medinipur, and Hooghly everywhere such vengeful criminal act scars the life of many young women just for rejection of love or marriage proposals, past revenge, domestic violence, and high dowry demands, and cheap and easy availability of acid in open markets. Thus, glimpses of newspaper records indeed envisage specific geospatial extent of differential crime in West Bengal.

Results of Spatial Autocorrelation (Moran's I)

The significant discernible spatial pattern of different forms of crime against women in West Bengal has evolved through the findings of Moran's I analysis (Fig. 3.6). From the perspective of crime mapping, spatial association statistics examine the existence of spatial autocorrelation between data variables indicating if positive or negative spatial autocorrelation exists. Henceforth, Moran's I offer additional statistical robustness to make core insight into the areal distribution of crime events and identify areas of high concentration of crime. The findings indicate that higher positive values of Moran's I (i.e., cruelties by husband or his relatives and dowry deaths) imply a more pronounced inclination toward spatial clustering, suggesting the presence of localized crime hotspots. And lower positive values suggest a relatively weaker disposition for spatial clustering of crime events in case of rape, kidnapping, and abduction. The Moran's I of cruelties by husband or his relatives (Sec.498-A IPC) indicates a notably higher value of 0.315 (z-score 2.75) (Part A of Fig. 3.6) in comparison to others positing a predisposition of strong positive geographical clustering. It implies that the districts with high occurrences of cruelties tend to be clustered geographically. Likewise, incidences of dowry deaths (Sec.304 B IPC) and

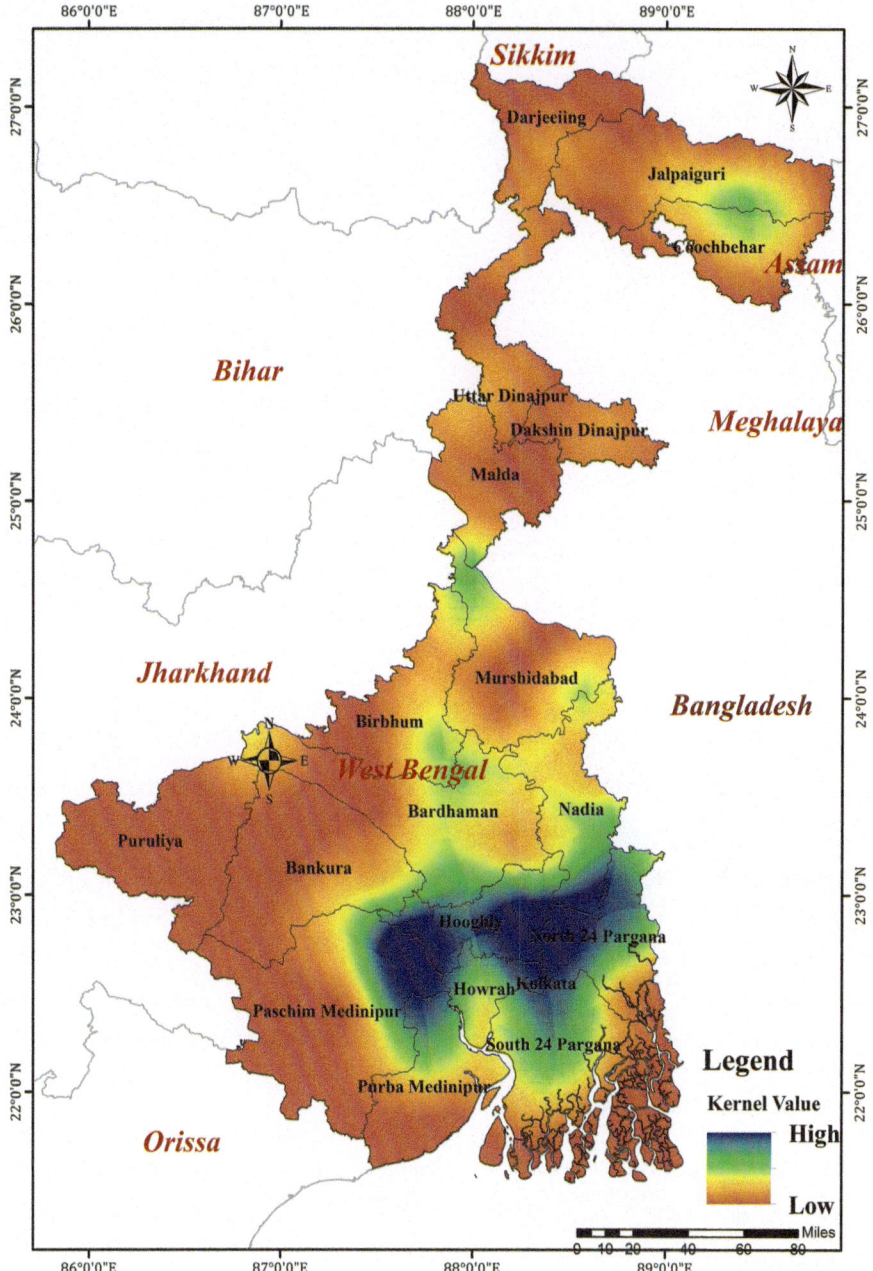

Fig. 3.5 Spatial extent of incidences of acid attacks against women in West Bengal. (Source: Newspaper reports (Times of India, The Telegraph, The Hindu, Anandabazar Patrika) and ASFI records. (Modified after Biswas & Chatterjee, 2018))

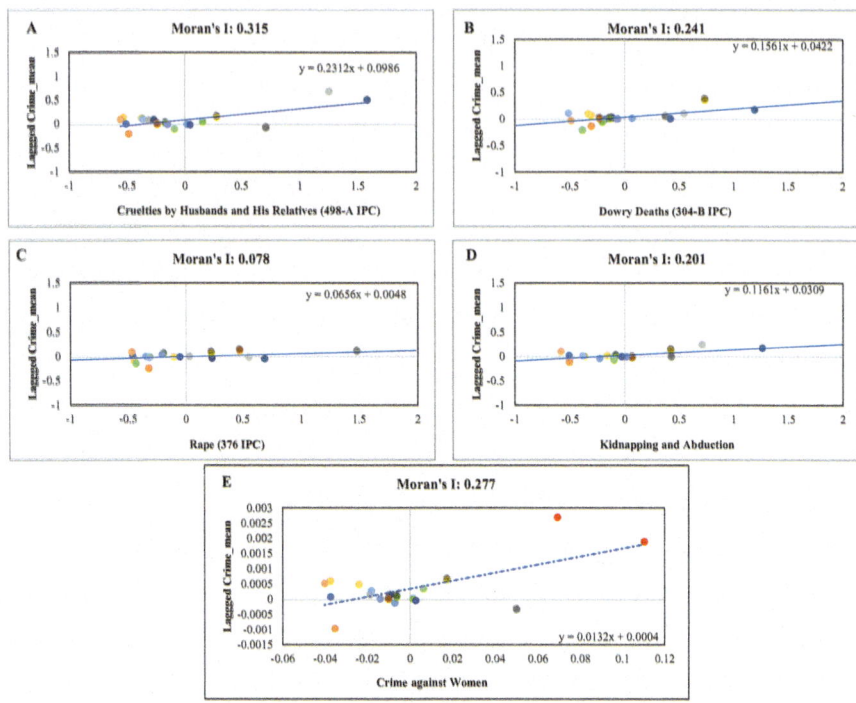

Fig. 3.6 Spatial autocorrelation (Moran's I) pattern of different forms of crime against women in West Bengal: (**a**) cruelties by husband or his relatives (sec. 498-A IPC), (**b**) dowry deaths (sec. 304 B IPC), (**c**) rape (sec. 376 IPC), (**d**) kidnaping and abduction (secs. 363–373 IPC), (**e**). overall crime against women. *$p < 0.01$. (Source: Author's computation)

Kidnapping and Abduction (Sec. 363–373 IPC) exhibit a substantial Moran's I value of 0.241 (z-score 2.15) and 0.201 (z-score 1.89) respectively (Part B and D of Fig. 3.6) indicating moderate spatial autocorrelation, implying that districts with elevated incidences of dowry deaths and kidnapping also exhibit a predisposition of geographical clustering. Whereas, Moran's I value of rape (Sec. 376 IPC) (Part C of Fig. 3.6) exhibits very minimal positive spatial associations or randomness ($I = 0.078$, z-score 1.12). It implies that the districts with higher instances of rape across West Bengal exhibit diminished propensities of geographical clustering and aligns closer to randomness. However, when considering all forms of crime against women (both IPC (Indian Penal Code) and CrPC (Criminal Procedure Code)), the corresponding Moran's I value ($I = 0.277$, z-score 2.99) indicates propensities of moderate positive spatial autocorrelation, suggesting significant geographic clustering patterns across West Bengal.

Apart from global spatial autocorrelation, the local Moran's I statistics assess the local spatial associations in the analyzed data and determine which neighborhoods posits statistically significant clusters. It yields fascinating valuable insights to distinguish areas of high concentration of crime and others marked by low crime concentration thereof. A significant spatial clustering pattern of crime against women in West Bengal has been evolved from the aggregated 12 years recorded data (Fig. 3.7a, b) which specifies that the districts Murshidabad, North and South-24 Parganas, Bardhaman, Nadia, Howrah, Hooghly, and Kolkata have highest concentration of overall incidences of crime against women. In contrast, Purulia, Bankura, Dakshin Dinajpur, and Birbhum experienced relatively lower concentrations of crimes against women during this period (Fig. 3.7a). Yet, the gross Moran's I statistics ($I = 0.277$) specified the existence of moderate positive spatial autocorrelation of crime across West Bengal indicating significant distinct geographic clustering patterns nonetheless from the statistical analysis of local Moran's I clear evidences of high and low values clusters of districts has been evolved which deviates from what would be expected under geographical randomness (Fig. 3.7b). The statistical analysis confirms the significance of the 95% confidence interval. The results specify that districts with a high concentration of crime, such as North- and South-24 Parganas and Murshidabad, form a highly clustered pattern and are surrounded by other districts with relatively high values. A distinct moderate clustering pattern has been observed in North Bengal and some segments of South Bengal. However, no

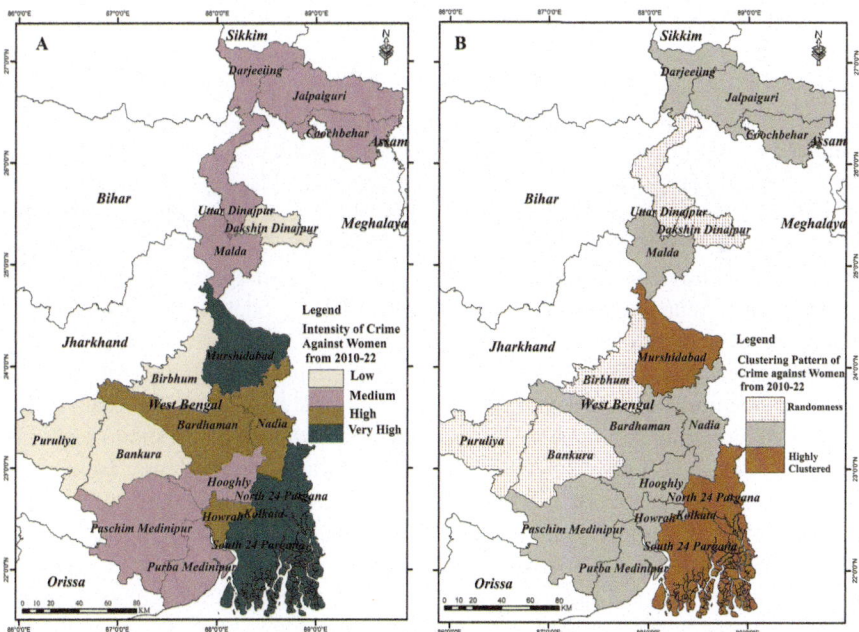

Fig. 3.7 Geographical clustering patterns of crime against women in West Bengal. (**a**) Spatial intensity across districts from 2010 to 2022, (**b**) high and low value clusters of districts, p value >0.05. (Source: data.gov.in; author's computation)

significant spatial clustering of women susceptibilities has been seen to be noticed in the districts of Uttar and Dakshin Dinajpur, Birbhum, Bankura, and Purulia or say these districts align closer to randomness concerning occurrences of crime against women during this period.

Crime Prediction with the ARIMA Model

Crime is a threat to justice and needs to be controlled immediately. To form a socially sustainable community and improve community safety and security, practitioners and policymakers instinctively focus on reducing crimes so that people can enjoy a quality life and social and economic prosperity can happen. Therefore, adopting improved mechanisms to analyze crime reports and forecast future crime trends is crucial to provide meaningful insights into crime phenomena and assist practitioners and policymakers in allocating additional resources for proactive measures to ensure public safety and security, leading to the emergence of a meaningful and sustainable society. Forecasting crime has gained importance in recent years as it directs investigative authorities in preventing criminal acts prior to their occurrence by adopting effective situational measures. Exact prediction and forecasting of crime is quite difficult, especially in unknown situations where predict the precise location and time of occurence of crime is uncertain and involves many computational challenges (Khashei & Bijari, 2011; Safat et al., 2021). Yet a precise estimation of the types of crime, crime rates, and future trends based on past patterns is necessary to ultimately support proactive policing strategies. Considerable research efforts and pilot projects were successful in the United States, Canada, South Korea, Chicago, and Los Angeles in predicting and forecasting crime using past crime datasets (Nasridinov & Park, 2014; Fitterer et al., 2015; Stec & Klabjan, 2018; Safat et al., 2021). Research groups around the world have proposed differential learning models to understand future crime trends, including the binary model, the threshold autoregressive model (TAR), long-short-term memory (LSTM), and other deep-learning approaches (Gamboa, 2017). Nonetheless, the literature notes the emerging challenges concerning the accuracy of crime predictions and forecasts (Nitta et al., 2019). Real-time forecasting of crime is always crucial, and crime data that shows seasonality reflects the real-time evolution of crime patterns in recent years. Time-series analysis incorporates diversified models, and ARIMA is considered one of the standard methods for time-series forecasting in recent research (Benabderrahmane et al., 2017; Zhang et al., 2020). ARIMA is a composite model in time-series forecasting comprising a traditional autoregressive moving average (ARMA), autoregressive (AR), and moving average (MA) (Siami-Namini & Namin, 2018). It computes temporal structures using a linear regression-based approach to accomplish one- or multistep-rolling forecasting (Safat et al., 2021). ARIMA analysis specifies the best predictive performances for the data of interest (Safat et al., 2021). This model favorably presents a distribution of the results obtained for each dataset, depending on the past window. For the effective applicability of this statistical model, the present study is going to forecast future crime trends in West Bengal for the next 5 years using ARIMA to support effective policing.

Model Architecture of ARIMA

Building an ARIMA model involves three iterative steps: a) identification, b) parameter estimation, and c) diagnostic checking. Before the implementation of the ARIMA model for forecasting crime, the datasets were preprocessed to reduce noise. Later, the Dickey-Fuller test was performed to scrutinize data stationarity. Next, nonlinear optimization procedures were applied to minimize the overall error rate. For that, a scale-dependent error measure, namely the root mean square error (RMSE), was calculated, along with the number of epochs and batch size to aggregate the magnitude of errors. The final step of the ARIMA model is diagnostic checking, which satisfies the model regarding error assumptions. These three steps are repeated until the desired model is built, which can then be used for predicting crime. A detailed structure of the ARIMA model architecture is illustrated in Fig. 3.8.

The ARIMA model has three components: (a) AR (the autoregressive part, which accounts for a series of dependence on its past value, "p"), (b) I (the integrated part, which accounts for the trend of component of a time-series, "d"), and (c) MA (the moving average part, which accounts for the random component of the time-series, "q"). The mathematical description of the ARIMA (p, d, q) model encompasses a variable's future value, which is assumed to be a linear function of several historical observations and random errors. The underlying mechanism uses the mean count to forecast in a time series. Based on rolling forecasting origin, the algorithm in ARIMA focuses on a single forecast and the next data point to predict. The mathematical equation of the ARIMA model is described below:

$$\varnothing(B)\nabla_d(yt-\mu)=\theta(B)_{et}$$

where yt is the actual value at time t. $\varnothing(B)=1-\sum_{i=1}^{p}\varphi_i B^i$, $\theta(B)=1-\sum_{j=1}^{q}\theta_j B^j$ are polynomials in B of degrees p and q. $\varnothing_i(i = 1, 2, 3, …. p)$ and $\theta_j(j = 1, 2, 3, ….q)$ are model parameters. p and q are integers and sometimes referred to as the order of the model. d is an integer and is often referred to as the order of differencing. $\Delta = (1 - B)$, B is referred to as a backward shift operator. et is considered random errors at time t and are assumed to be independently distributed with a mean of 0 and a constant variance of σ^2.

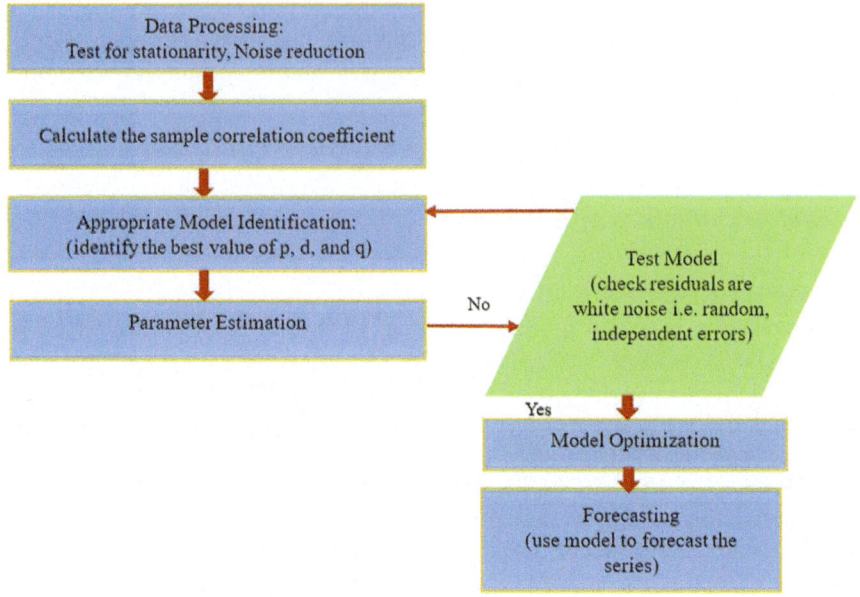

Fig. 3.8 Architecture of the ARIMA model

Results of ARIMA

The application of the ARIMA model demonstrated the best predictive performance for the extensive visualization of crime particulars and forecasted crime to support proactive policing. This statistical analysis embodied the distribution of results obtained for each dataset with all architectures based on the past window length and forecasted crime to support proactive policing. The prediction result of the ARIMA (1, 3, 4) analysis for assessing the next 5-year trend of crimes against women in West Bengal is represented in Fig. 3.9. The mean crime count is calculated to estimate the future crime trend. The parameters p, d, and q are selected after a glance at the ACF (autocorrelation function) and PCAF (partial autocorrelation function) plots. The summary of the model is presented in Table 3.1. The stationary R-squared value represents the goodness of fit of the model, indicating that nearly 78 percent of the total variance in the series is explained by this model. The values of RMSE (834.520), MAE (mean absolute error) (675.341), and normalized BIC (21.524) are within the satisfactory ranges, which denotes that the model is a good fit. The p-value of the Ljung Box statistics (Q) lies above 0.05, indicating that the residuals are white noise, which means the residuals are completely independent and that the model can be accepted. The ARIMA (1, 3, 4) result shows that the forecasted values of crime against women seem to be synchronized with past crime values, indicating significant variation in recent years. Since 2014, the rate of incidences had followed a downward trend in West Bengal, but it again followed a moderately upward pattern from 2019 to 2022, and this variation is expected to follow an upward rising trend in the near future (Fig. 3.9).

Discussions and Conclusion

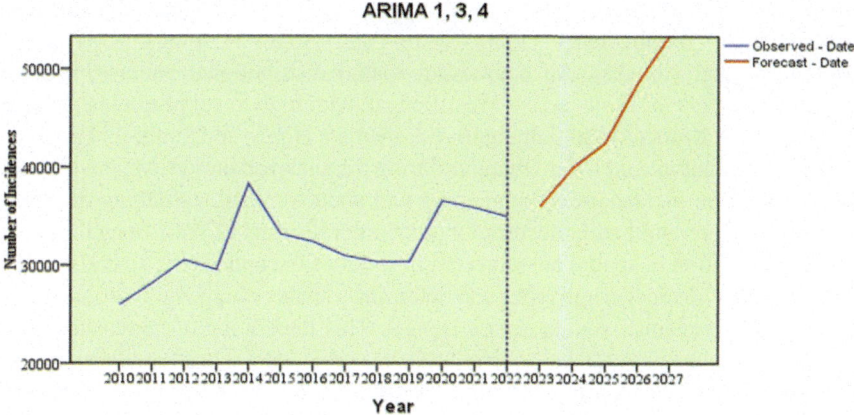

Fig. 3.9 Forecasting crime against women in West Bengal using ARIMA (1, 3, 4)

Table 3.1 Model statistics

Model	Number of predictors	Model fit statistics				Ljung-Box Q(18)			Number of outliers
		Stationary R-squared	RMSE	MAE	Normalized BIC	Statistics	DF	Sig.	
Crime against women-model_1	0	0.775	834.520	675.341	21.524	280.112	12	0.167	0

Discussions and Conclusion

The present chapter provides an insightful understanding of the spatiotemporal evolvement pattern of crime against women in West Bengal from multivariate perspectives, with a specific focus on four major crime categories experienced by women, and ascertains the future crime trends to endorse space-based policing. The integration of statistical methods with GIS tools is indispensable for the comprehensive understanding of crime patterns and the spatial modeling of crime data. The findings unveil a concerning upward trajectory of the incidents of crime against women in West Bengal during the analyzed period from 2010 to 2022. Notably, the districts of North- and South-24 Parganas, Malda, Murshidabad, Bardhaman, Nadia, Howrah, Kolkata, Hooghly, Cooch Behar, and Jalpaiguri have emerged as crime hotbeds with alarmingly high instances of crime against women. In contrast, the districts of Darjeeling, Uttar and Dakshin Dinajpur, Bankura, Purulia, and Birbhum have experienced comparatively fewer instances of crimes perpetrated against women during the analyzed period.

More precisely, the border districts of West Bengal are at high risk when it comes to the overall concentration of crimes against women. Apart from this, distinct spatial extension patterns for different crime categories have evolved from periodic data insights across West Bengal. Cruelties by husbands or their relatives and incidents of dowry deaths have predominantly been observed in the districts of Murshidabad, Bardhaman, Nadia, North- and South-24 Parganas, Paschim Medinipur, and Hooghly. An engrossing spatial distributional pattern has been observed for rape incidents across West Bengal, with high to very high instances in the districts of Murshidabad, Jalpaiguri, Bardhaman, North- and South-24 Parganas, Paschim Medinipur, and Uttar Dinajpur during the analyzed period. Meanwhile, the highest average incidences of kidnapping and abduction and trafficking of young women have predominantly occurred in the border districts of West Bengal.

The estimation of spatial autocorrelation signifies the underlying spatial clustering patterns of distinct crime types and determines areas of high and low concentrations of crime against women across Bengal. This integration of geovisualization techniques alongside spatial autocorrelation has provided a robust understanding of the spatiotemporal dynamism of crime against women across West Bengal as well as unveiled underlying clustering patterns at large. Additionally, the implementation of a deep-learning model architecture for time-series analysis through ARIMA to forecast 5-year crime trends signifies an upward variation in crimes against women in West Bengal in the future. This model can also be applied to forecast individual crime trends to assist law enforcement authorities in taking situational prevention measures to address specific crime issues and reduce women's susceptibilities.

Overall, all these results unveil the spatiotemporal dynamism of crime against women in West Bengal, identify the most vulnerable districts with higher crime rates, signify spatial clustering patterns, and determine future crime trends with improved predictive accuracy. This study could be beneficial for practitioners and law enforcement authorities by providing meaningful insights into crime occurrences against women and supporting legitimate crime prevention measures to ensure women's safety throughout West Bengal. For future work, this study will be extended through the creation of different hybrid models to enhance crime prediction accuracy and improve overall performance.

Appendices

Appendix 1: Summarization of Some Published Newspaper Reports on Women Trafficking in North- and South-24 Parganas Districts, West Bengal

Appendices

Source	Rescue from	Destination	Path	Age of victim	Offender's identity	Year	Reason	No. of victims	Under police station	District	Hyperlink
Bhangor	Delhi	Delhi	Howrah	19	Rishi Raj (trafficker)	2017	On the pretext of	1	Bhangor PS	South-24 Pargana	https://timesofindia.indiatimes.com/city/kolkata/19-year-old-trafficked-girl-rescued-from-delhi/articleshow/57190533.cms
Bansberia		Chapra, Bihar	Howrah	15	Neighbors	2015	Dancing	1	Mograhat PS	South-24 Pargana	https://www.anandabazar.com/district/howrah-hoogly/bansberia-girl-rescued-from-chhapra-bihar-1.160273
Baruipur	Gajiabad	Ghaziabad, Delhi	Howrah	22	Lover	2015	Promise of marriage	1	Baruipur PS	South-24 Pargana	https://www.anandabazar.com/state/youth-sold-lady-love-to-women-traffickers-by-1-5-lakh-rupees-later-rescued-by-cid-1.219999
Bangladesh	Bandar area	Mumbai	Bashirhat	15	Rubel Dewan (trafficker)	2016	Promise of marriag	1	Habra, bandar PS	South-24 Pargana	https://www.anandabazar.com/district/24-paraganas/women-trafficking-situation-getting-worse-day-by-day-at-bongaon-1.480402
Canning (taldir)		UP (Khusinagar)	Howrah		Payel Haldar and Biswajit Haldar (Trafficker)	2017		1	Canning PS	South-24 Pargana	https://www.anandabazar.com/district/24-paraganas/4-girls-rescued-from-trafficking-1.523305
Joynagar		Haryana (Rohtak area)			Husband and a relative			1	Canning PS, and Joynagar PS	South-24 Pargana	https://www.anandabazar.com/district/24-paraganas/4-girls-rescued-from-trafficking-1.523305
Jibontala		Delhi (Rajouri garden)	Howrah		Neighbors		Promise of job	1	Jibantala PS	South-24 Pargana	https://www.anandabazar.com/district/24-paraganas/4-girls-rescued-from-trafficking-1.523305

Source	Rescue from	Destination	Path	Age of victim	Offender's identity	Year	Reason	No. of victims	Under police station	District	Hyperlink
Mathurapur	Agra	Agra	Howrah	15	Lover	2017	Marriage proposal	1	Mathurapur	South-24 Pargana	https://www.anandabazar.com/district/24-paraganas/charges-filed-against-human-trafficking-1.622983
Canning		Delhi			Mother	2016	Mother sold girl	1	Canning PS	South-24 Pargana	https://www.anandabazar.com/district/24-paraganas/woman-arrested-for-girl-trafficking-at-canning-1.494690
Sagar	Gobindapuri	Gobindapuri, Delhi			Lover	2017	Marriage	1	Sagar PS and basanti PS	South-24 Parganas	https://www.anandabazar.com/district/24-paraganas/girl-saved-before-trafficking-1.542517
Canning	Pune	Pune	Howrah via Sealdh	Unknown	Neighbours	2017	Prostitution	1	Canning PS	South-24 Parganas	https://www.anandabazar.com/state/another-ayesha-from-bengal-with-serious-injury-identified-from-delhi-1.606004
Maheshtala PS	Delhi	Delhi	Sealdah	15	Neighbours	2017	Marriage traffic	1	Maheshtala PS	South-24 Parganas	https://www.anandabazar.com/state/threat-of-trafficking-again-the-lady-is-fighting-who-saved-her-life-from-being-sold-1.582931
Diamond harbor	Delhi G.B.Road brothel	Delhi G.B. Road	Howrah	15	Unknown	2015	Prostitution	1	Diamond harbour	South-24 Parganas	https://www.anandabazar.com/national/police-took-class-at-school-to-stop-the-girl-s-kidnapping-1.537602
Joynagar	Bihar, Siwan area	Siwan, Bihar	Howrah	16	Neighbour	2017	Prostitution	1	Joynagar PS	South-24 Parganas	https://www.anandabazar.com/state/sex-workers-choose-to-finish-their-life-as-their-rehabilitation-does-not-happen-properly-1.712374

Appendices

Source	Rescue from	Destination	Path	Age of victim	Offender's identity	Year	Reason	No. of victims	Under police station	District	Hyperlink
Joynagar	Pune	Pune	Howrah	16	Monirul Molla (Neighbour)	2017	Dancing	1	Joynagar PS	South-24 Parganas	https://www.anandabazar.com/state/teenage-student-rescued-after-getting-kidnapped-but-still-child-traffickers-are-threatening-her-1.676567
Maheshtala	Pune	Pune	Howrah	15	Neighbours	2017	Prostitution	1	Maheshtala PS	South-24 Parganas	https://www.anandabazar.com/state/threat-of-trafficking-again-the-lady-is-fighting-who-saved-her-life-from-being-sold-1.582931
Nafarganj, basanti	Gobindapuri, Delhi	Gobindapuri, Delhi	Howrah via Sealdah	13	Unknown	2016	Prostitution	1	Canning PS	South-24 Parganas	https://www.anandabazar.com/district/24-paraganas/women-trafficking-from-sunderban-is-increasing-1.481071
Joynagar	Mumbai	Mumbai	Howrah	25		2016	Prostitution	1	Joynagar	South-24 Parganas	https://www.anandabazar.com/district/24-paraganas/women-trafficking-from-sunderban-is-increasing-1.481071
Sundarban	Mumbai	Mumbai	Howrah	21	Lover	2016	Prostitution	1	Sundarban PS	South-24 Parganas	https://www.anandabazar.com/district/24-paraganas/women-trafficking-from-sunderban-is-increasing-1.481071

Source: The Hindu, Times of India, The Telegraph, Anandabazar Patrika, Ei-Samay

Appendix 2: Summarization of Some Published Newspaper Reports on the Incidences of Acid Attacks Against Women in West Bengal

Place	Date	Year	No. of female victims	Victims' age	No. of perpetrators	Reason for attacks	Hyperlink
Khamedda village, Tarapith, Rampurhat, Birbhum	11-10-2017	2017	1	11–20 years	2	Refusal of love/sex proposal	https://www.anandabazar.com/district/purulia-birbhum-bankura/complaint-against-two-young-men-for-throwing-acid-to-a-college-student-1.687878
Daspur bus stand road, Ghatal, Paschim Medinipur	15-06-2016	2016	1	11–20 years	1	Refusal of love/sex proposal	https://www.anandabazar.com/state/acid-attack-on-a-college-girl-in-ghatal-1.411326
Slum area near Birati station	27-08-2016	2016	1	21–30 years	1	Refusal of love/sex proposal	https://www.anandabazar.com/state/demonetisation-gives-more-trouble-to-this-acid-attacked-family-1.530809
Sabang, Paschim Medinipur	23-02-2017	2017	1	11–20 years	1	Other	https://www.anandabazar.com/state/examinee-can-t-appeared-in-exam-1.568864
Nowda, near Chandpur Panchayet, Domkal	18-08-2016	2016	1	11–20 years	1	Refusal of love/sex proposal	https://www.anandabazar.com/district/nadia-murshidabad/experience-shared-by-an-acid-attack-victim-1.577233
Bagda, Chormondal village	10-04-2010	2010	1	31–40 years	1	Marital dispute	https://www.anandabazar.com/state/father-of-acid-attack-victim-sell-land-for-daughter-s-medical-treatment-1.260287
Humgarh more, Goaltore, Paschim Medinipur	28-08-2016	2015	1	> 40 years	1	Other	https://www.anandabazar.com/district/midnapore/fruit-seller-arrested-for-acid-attack-1.197376
Usti, Karbala Bazaar, S-24 Parganas	02-06-2017	2017	2	11–20 years	1	Refusal of love/sex proposal	https://www.anandabazar.com/district/24-paraganas/golapi-khatun-create-a-new-record-1.621996
Haripal, Hugly, Dullya village of Narayanpur Bahirkhand Panchayet	07-05-2015	2015	1	11–20 years	1	Refusal of love/sex proposal	https://www.anandabazar.com/district/howrah-hoogly/haripal-youth-arrested-for-acid-attack-on-school-student-1.142610

Appendices

Place	Date	Year	No. of female victims	Victims' age	No. of perpetrators	Reason for attacks	Hyperlink
Sialdah-Kalyanpur Diamond Harbor local	12-02-2017	2017	2	21–30 years and 31–40 years	1	Other	https://www.anandabazar.com/state/hospital-writes-chemical-attacked-instead-of-acid-attacked-in-discharge-certificate-of-a-lady-1.568398
Kayaradanga, Khaspara, Katowa, Bardwan	02-08-2017	2017	1	31–40 years	6	Domestic violence/ family related dispute	https://www.anandabazar.com/state/husband-allegedly-attack-wife-with-a-sharp-weapon-and-acid-for-giving-birth-a-daughter-1.651648
Bankimnagar, Dhantala, Ranaghat, Nadia	09-11-2016	2016	1	21–30 years	1	Marital dispute	https://www.anandabazar.com/district/nadia-murshidabad/husband-arrested-for-throwing-acid-at-wife-cost-only-rs15-1.510425
Cooch Behar, B.N.Shill College	15-09-2015	2015	1	11–20 years	1	Refusal of love/sex proposal	https://www.anandabazar.com/district/north-bengal/lifetime-imprisonment-for-a-convict-on-acid-attack-case-1.611916
Nandanpur, Daspur, Paschim Medinipur	30-05-2015	2015	1	11–20 years	1	Refusal of love/sex proposal	https://www.anandabazar.com/eimuhurte/acid-attack-on-a-student-at-midnapore-arrested-two-1.152875
Hanskhali, Gajna, Dakshin Para, Nadia	15-10-2016	2016	1	11–20 years	1	Refusal of love/sex proposal	https://www.anandabazar.com/district/nadia-murshidabad/member-of-save-democracy-form-visits-deceased-acid-attacks-family-1.498986
Baishabnagar, Satangapara, Maldah	19-06-2014	2014	1	11–20 years	3	Refusal of love/sex proposal	https://www.anandabazar.com/district/north-bengal/miserable-life-of-an-acid-attacked-victim-1.477534
Chanchal, Harischrandrapur, Maldah	30-08-2015	2015	1	31–40 years	1	Marital dispute	https://www.anandabazar.com/district/north-bengal/monetary-compensation-for-the-victim-of-acid-attack-1.483682
Gajna, Hanskhali, Nadia	10-10-2016	2016	1	11–20 years & 31–40 years	1	Rejection of marriage proposal	https://www.anandabazar.com/state/mother-daughter-was-wounded-by-acid-attack-1.493651
Alipurduwar, Udaybitan area, Suryanagar	05-01-2017	2017	1	> 40 years	1	Other	https://www.anandabazar.com/district/north-bengal/mystery-creates-about-a-acid-attack-on-a-elederly-1.543641
Nikasi area, Moyna, Tamluk, Purba Medinipur	09-01-2017	2017	1	> 40 years	2	Refusal of love/sex proposal	https://www.anandabazar.com/district/midnapore/one-man-arrested-for-attacking-woman-with-acid-1.671771

Place	Date	Year	No. of female victims	Victims' age	No. of perpetrators	Reason for attacks	Hyperlink
Kholta, under Kotowali P.A, Cooch Behar	06-10-2015	2015	1	11–20 years	2	Refusal of love/sex proposal	
Dholhat, Gopinathpur, S-24 Parganas	05-06-2017	2017	1	21–30 years	Not known	Other	https://www.anandabazar.com/district/24-paraganas/some-unknown-person-were-accused-of-throwing-acid-on-a-housewife-1.623473
Rishra, Hugly	12-14-2014	2014	1	31–40 years	1	Other	https://www.anandabazar.com/district/howrah-hoogly/an-acid-attacked-women-did-not-get-compensation-from-government-1.549483
Polba, Mahanad, Hugly	15-09-2013	2013	1	> 40 years	1	Other	https://www.anandabazar.com/district/howrah-hoogly/again-an-acid-attack-in-hooghly-compensation-proposal-been-made-1.573850
Dhanekhali, Rojipur, Kaknadi area, Hugly	06-07-2014	2014	1	11–20 years	1	Refusal of love/sex proposal	https://www.anandabazar.com/district/howrah-hoogly/again-an-acid-attack-in-hooghly-compensation-proposal-been-made-1.573850
Sarbamongla Para, Bardwan	25-03-2017	2017	1	> 40 years	2	Other	https://www.anandabazar.com/district/bardhaman/again-an-acid-attack-in-district-1.585782
Ghatal, Paschim Medinipur	26-02-2017	2017	1	> 40 years	1	Rejection of marriage proposal	https://www.anandabazar.com/state/youth-attacked-his-girlfriend-with-acid-1.542488
Joynagar 6 no. Ward, Bardwan, Kolkata	14-07-2016	2016	1	11–20 years	6	Refusal of love/sex proposal	https://www.anandabazar.com/state/acid-attack-victim-manisha-determined-to-rebuild-her-life-1.434407
Dampal, Purbasthali, Bardwan	12-12-2016	2016	1	11–20 years	1	Refusal of love/sex proposal	https://www.anandabazar.com/state/acid-attack-on-a-teenage-girl-while-sleeping-with-her-mother-1.530799
Daria, Canning, S-24 Parganas	06-04-2015	2015	1	11–20 years	1	Refusal of love/sex proposal	https://www.anandabazar.com/district/24-paraganas/acid-attack-on-a-student-arrested-one-1.131435
Tarakeswar, Piyasara, Kolonipara, Hugly	25-07-2016	2016	1	> 40 years	1	Other	https://www.anandabazar.com/state/acid-attack-by-neighbour-woman-and-2-sons-injured-dgtl-1.442251

Appendices

Place	Date	Year	No. of female victims	Victims' age	No. of perpetrators	Reason for attacks	Hyperlink
Naranda, Panskura, Purba Medinipur	08-07-2014	2014	1	11–20 years, 21–30 years and > 40 years	1	Rejection of marriage proposal	https://www.anandabazar.com/state/acid-attack-at-tamluk-1.47234
Muchlandapur station, N-24 Parganas	30-11-2010	2010	1	11–20 years	1	Refusal of love/sex proposal	https://www.anandabazar.com/state/accused-gets-lifetime-imprisonment-for-acid-attack-1.486711
Baruipur station, in the train, South-24 Parganas	14-02-2017	2017	2	31–40 years	1	Land/property/money dispute	https://www.anandabazar.com/district/24-paraganas/woman-injured-for-acid-attack-in-train-1.564055
Dhupguri 6 no Ward Vibekananda Para, Jalpaiguri	05-03-2017	2017	1	11–20 years	1	Refusal of love/sex proposal	https://www.anandabazar.com/district/north-bengal/woman-locked-herself-in-house-in-the-fear-of-acid-attack-1.574302
Karimpur bazaar, Murshidabad	29-11-2014	2014	1	31–40 years	1	Marital dispute	https://www.anandabazar.com/district/nadia-murshidabad/%E0%A6%85%E0%A6%AD-%E0%A6%AF-%E0%A6%95-%E0%A6%A4-%E0%A6%85%E0%A6%A7%E0%A6%B0-%E0%A6%89%E0%A6%A6-%E0%A6%AC-%E0%A6%97-%E0%A6%A8-%E0%A6%95%E0%A6%B0-%E0%A6%AE%E0%A6%AA-%E0%A6%B0-1.91193
Dabra village, Pandua, Chuchura, Hugly	25-08-2012	2012	1	21–30 years	2	Refusal of love/sex proposal	https://www.anandabazar.com/district/24-paraganas/%E0%A6%85-%E0%A6%AF-%E0%A6%B8-%E0%A6%A1-%E0%A6%9B-%E0%A7%9C-%E0%A6%AF-%E0%A6%AC%E0%A6%9C-%E0%A6%9C-%E0%A6%AC%E0%A6%A8-1.26171
Arambag, Hugly/Joypur, Bankura	28-07-2014	2014	1	11–20 years	4	Refusal of love/sex proposal	https://www.anandabazar.com/eimuhurte/%E0%A6%86%E0%A6%B0-%E0%A6%AE%E0%A6%AC-%E0%A6%97-%E0%A6%85-%E0%A6%AF-%E0%A6%B8-%E0%A6%A1-%E0%A6%B9-%E0%A6%AE%E0%A6%B2-%E0%A6%95%E0%A6%B2-%E0%A6%9C-%E0%A6%9B-%E0%A6%A4-%E0%A6%B0-%E0%A6%95-1.54467

Place	Date	Year	No. of female victims	Victims' age	No. of perpetrators	Reason for attacks	Hyperlink
Atkula village, Nalhati, Birbhun	12-08-2014	2014	1	31–40 years	1	Marital dispute	https://www.anandabazar.com/district/purulia-birbhum-bankura/ঘ-মন-ত-মহ-ল-র-গ-য়-স-ড-ধ-ত-স-ব-ম-1.58643
Diyara, Ramchandrapur, Singur, Hugly	31-12-2014	2014	1	11–20 years	1	Other	https://www.anandabazar.com/district/howrah-hoogly/তর-গ-র-উপর-অ-য-স-ড-হ-মল-য়-দ-য-ক-ত-অধর-ই-1.101184
Duars, Falakata Madari Road, Alipurduwar	18-05-2014	2014	1	11–20 years	2	Rejection of marriage proposal	https://www.anandabazar.com/district/north-bengal/প-র-ম-প-রত-য-খ-য-ত-হয়-ক-শ-র-র-ম-খ-অ-য-স-ড-1.32729
Kulti, Chittaranjan station, Bardwan district	08-06-2014	2014	1	31–40 years	1	Rejection of marriage proposal	https://www.anandabazar.com/eimuhurte/প-র-ম-প-রত-য-খ-য-ন-তর-গ-র-br-ম-খ-অ-য-স-ড-প-র-ক-তন-প-র-ম-ক-র-1.41896
Jalangi, Murshidabad, Domkol	07-11-2014	2014	2	Below 10 years & 11–20 years	2	Refusal of love/sex proposal	https://www.anandabazar.com/eimuhurte/প-র-ম-ব-যর-থ-হয়াক-শ-র-র-উপর-অ-য-স-ড-হ-মল-ধ-ত-২-1.84638
Parapukur G.N. Mitra Lane, Bardwan	25-09-2014	2014	1	11–20 years	1	Refusal of love/sex proposal	https://www.anandabazar.com/district/bardhaman/ব-ড-ত-চ-ক-কল-জ-ছ-ত-র-র-ম-খ-অ-য-স-ড-1.53588
Islampur, under Chopra police station; Kaligunj; Chutiakhor Panchayet; Uttardinajpur	09-07-2014	2014	1	> 40 years	1	Other	https://www.anandabazar.com/district/north-bengal/ব-দ-ধ-দম-পত-ক-অ-য-স-ড-প-উ-য়-উধ-ও-দ-ই-দ-য-ক-ত-1.48647
Rishra, Panchlaki area, Baruapara, Hugly	06-12-2014	2014	1	31–40 years	1	Refusal of love/sex proposal	https://www.anandabazar.com/eimuhurte/য-বত-র-ম-খ-অ-য-স-ড-ছ-ড-আত-মহত-য-র-চ-য-ট-1.93630
Raygunj, Sebak Palli, North Dinajpur, South Dinajpur Harirampur	18-01-2014	2014	1	11–20 years	1	Other	https://www.anandabazar.com/district/north-bengal/র-য়গঞ-জ-অ-য-স-ড-হ-মল-র-শ-ক-র-ছ-ত-র-1.106496
Habra, N-24 Parganas	27-07-2014	2014	1	31–40 years	1	Marital dispute	https://www.anandabazar.com/district/24-paraganas/স-ত-র-ক-অ-য-স-ড-ছ-ড-ধ-ত-স-ব-ম-জ-ল-ব-ক-র-ত-ও-1.63044
Gopalpur village, Bolpur, Birbhum	24-10-2011	2011	3	Below 10 years, 31–40 years & > 40 years	1	Marital dispute	https://www.anandabazar.com/district/purulia-birbhum-bankura/স-ত-র-ক-অ-য-স-ড-ছ-ড-য়-দ-য-স-ব-যস-ত-স-ব-ম-1.55172

Appendices

Place	Date	Year	No. of female victims	Victims' age	No. of perpetrators	Reason for attacks	Hyperlink
Sebak road, Siliguri, Jalpaiguri (38%), and Darjeeling (62%)	22-12-2014	2014	2	21–30 years	2	Rejection of marriage proposal	https://www.anandabazar.com/district/north-bengal/অ-য-স-ড-ক-গ-ড-গ-র-ফত-র-আরও-এক-জন-1.99851
Joynagar, Mojilpur 5 no Ward, Bardwan, Kolkata	19-08-2016	2016	1	31–40 years	1	Domestic violence/ family related dispute	https://www.anandabazar.com/district/24-paraganas/woman-accused-of-acid-attack-on-woman-1.459447
S-24 Parganas Taldi, Canning, Chandkhali Area	22-12-2016	2016	2	11–20 years	1	Refusal of love/sex proposal	https://www.anandabazar.com/district/24-paraganas/two-schoolgirls-face-was-scorched-by-acid-attack-1.535666
Patuli, Purba Burdwan	18-05-2015	2015	1	> 40 years	1	Land/ property/ money dispute	https://www.anandabazar.com/calcutta/acid-attack-on-family-members-injured-one-dgtl-1.390062
Gaighata, N-24 Parganas	–		1	31–40 years	1	Marital dispute	–
Chondrakona village, Ghatal, Paschim Medinipur	–		1	31–40 years	2	Domestic violence/ family related dispute	https://www.anandabazar.com/district/midnapore/acid-thrown-to-a-woman-for-not-withdrawing-case-2-arrested-1.566768
Mahisa PS, Habra, N-24 Parganas	03-10-2017	2017	1	31–40 years	1	Refusal of love/sex proposal	https://www.anandabazar.com/district/24-paraganas/man-arrested-for-illegal-trading-of-acids-1.683203
Ranaghat	–	2014	1	21–30 years	Not known	Land/ property/ money dispute	–
Birbhum	–	2014	1	Below 10 years	Not known	Land/ property/ money dispute	–
Tamluk	–	2014	1	Below 10 years	Not known	Land/ property/ money dispute	–
Rajarhat	–	2014	1	Below 10 years	Not known	Land/ property/ money dispute	–
Salt Lake	–	2013	1	31–40 years	Not known	Marital dispute	–
Khardah	–	2013	1	21–30 years	Not known	Not known	–

Place	Date	Year	No. of female victims	Victims' age	No. of perpetrators	Reason for attacks	Hyperlink
Diamond harbour	–	2012	1	11–20 years	Not known	Land/property/money dispute	–
Kalyani	–	2012	1	11–20 years	Not known	Land/property/money dispute	–
Malda	–	2012	1	11–20 years	Not known	Not known	–
Howrah	–	2011	1	21–30 years	Not known	Not known	–
Kolkata	–	2010	1	11–20 years	Not known	Sadistic pleasure	–
Birballabpara	–	2010	1	11–20 years	Not known	Eve teasing	–
Harua, Murshidabad, Po Debpur, PS Suti	26-09-2011	2011	1	11–20 years	1	Other	–
Kartiya bazaar, near Kartiya Durga Mandap, Po Nonakuri bazar, PS Tamluk, Purba Medinipur	12-08-2013	2013	2	Below 10 years & 11–20 years	1	Refusal of love/sex proposal	–
Nandamar, Purba Medinipur	15-07-2014	2014	1	11–20 years	2	Refusal of love/sex proposal	–
Swarupnagar, S-24 Parganas	13-01-2014	2014	1	> 40 years	2	Other	–
	23-06-2012	2012	1	31–40 years	Not known	Not known	–
Howrah Moidan Moti Ghosh lane	09-08-2017	2017	Male victim (1)	21–30 years	1	Other	https://www.anandabazar.com/calcutta/2-injured-by-acid-attack-1.655190
Balurghat Mallikpur, South Dinajpur	18-05-2016	2016	Female victim (1)	11–20 years	3	Land/property/money dispute	–
Chondrakona PS, Khirpai, Gancha village, Ghatal, Paschim Medinipur	29-06-2017	2017	Male victim (1)	21–30 years	2	Other	https://www.anandabazar.com/district/midnapore/acid-attack-on-tmc-worker-1.635188
Daspur 1 Block, Nandanpur Panchayet, Paschim Medinipur	20-03-2004	2004	Male victim (1)	> 40 years	12	Other	https://www.anandabazar.com/district/midnapore/acid-attack-life-term-imprisonment-to-seven-accused-1.566224

Place	Date	Year	No. of female victims	Victims' age	No. of perpetrators	Reason for attacks	Hyperlink
Hazinagar Marawari Kol, Naihati and Halisahar Er Majhe, N-24 Parganas	11-09-2016	2016	Male victim (1)	31–40 years	5	Other	https://www.anandabazar.com/state/acid-attacks-to-an-elderly-in-syndicate-violence-1.475601
Paschim Medinipur, Ghatal, Kusman village	26-05-2017	2017	Male victim (1)	21–30 years	2	Other	https://www.anandabazar.com/state/acid-thrown-at-youth-in-ghatal-1.618421
Bankura, Joypur, Chotra village	16-03-2017	2017	Male victim (1)	Below 10 years	2	Land/property/money dispute	https://www.anandabazar.com/district/purulia-birbhum-bankura/neighbour-throws-acid-to-a-child-for-family-broil-1.580428
Gopalnagar, Battala, Narapukur area, Banga, N-24 Parganas	03-07-2017	2017	Male victim (1)	21–30 years	2	Other	https://www.anandabazar.com/district/24-paraganas/youth-trying-to-kill-a-man-with-acid-1.636994
Ghhatal PS, Balidanga village	05-06-2017	2017	Male victim (1)	11–20 years	2	Not known	https://www.anandabazar.com/district/midnapore/hanging-body-found-of-main-accused-of-ghatal-acid-attack-1.623365
Kulti, Rathanagar Cenemahal, slums, Paschim Burdwan	23-09-2017	2017	Male victim (1)	> 40 years	10	Other	https://www.anandabazar.com/district/bardhaman/acid-attack-on-a-middle-aged-man-for-registered-protest-against-liquor-consumption-1.679239
Naya Bahadurpur, Suti 1 Murshidabad	07-06-2012	2012	Female victim (1)	21–30 years	2	Other	–

Source: The Times of India, The Telegraph, The Hindu, Anandabazar Patrika, and ASFI Records

References

Baller, R. D., Anselin, L., Messner, S. F., Deane, G., & Hawkins, D. F. (2001). Structural covariates of US county homicide rates: Incorporating spatial effects. *Criminology, 39*(3), 561–588.

Benabderrahmane, S., Mellouli, N., Lamolle, M., & Paroubek, P. (2017). Smart4job: A big data framework for intelligent job offers broadcasting using time series forecasting and semantic classification. *Big Data Research, 7*, 16–30.

Biswas, P., & Chatterjee, N. D. (2018). Acid attacks: A threat against women- measure the effectiveness of existing legislatures to curb acid violence. *Eastern Geographer, XXIV*(1), 149–158.

Eck, J., Chainey, S., Cameron, J., & Wilson, R. (2005). *Mapping crime: Understanding hotspots.* National Institute of Justice.

Fitterer, J., Nelson, T. A., & Nathoo, F. (2015). Predictive crime mapping. *Police Practice and Research, 16*(2), 121–135.

Gamboa, J. C. B. (2017). Deep learning for time-series analysis. arXiv preprint arXiv:1701.01887.

Khashei, M., & Bijari, M. (2011). A novel hybridization of artificial neural networks and ARIMA models for time series forecasting. *Applied Soft Computing, 11*(2), 2664–2675.

Nasridinov, A., & Park, Y. H. (2014). A study on performance evaluation of machine learning algorithms for crime dataset. *Advanced Science and Technology Letters (Networking Communication 2014), 66*, 90–92.

Nitta, G. R., Rao, B. Y., Sravani, T., Ramakrishiah, N., & BalaAnand, M. (2019). LASSO-based feature selection and naïve Bayes classifier for crime prediction and its type. *Service Oriented Computing and Applications, 13*, 187–197.

Safat, W., Asghar, S., & Gillani, S. A. (2021). Empirical analysis for crime prediction and forecasting using machine learning and deep learning techniques. *IEEE Access, 9*, 70080–70094.

Siami-Namini, S., & Namin, A. S. (2018). Forecasting economics and financial time series: ARIMA vs. LSTM. arXiv preprint arXiv:1803.06386.

Stec, A., & Klabjan, D. (2018). Forecasting crime with deep learning. arXiv preprint arXiv:1806.01486.

Zhang, Z., Sha, D., Dong, B., Ruan, S., Qiu, A., Li, Y., et al. (2020). Spatiotemporal patterns and driving factors on crime changing during black lives matter protests. *ISPRS International Journal of Geo-Information, 9*(11), 640.

Chapter 4
Predicting Future Crime Hotspots Using Statistical Techniques and GIS

Introduction

Criminality has long been a matter of great concern at both the national and local levels. It is one of the causes of social disorders, hardships, and misfortunes to the social and economic environment. Increasing crime pose immense challenges to law enforcement agencies to initiate targeted interventions and employ resources to safeguard people (GOSCHIN, 2019). So, understanding the causes of criminality is indispensable in formulating effective measures to combat criminogenic events. Over the past two decades, the Geographic Information System (GIS) has been intensively used in academia and law enforcement agencies to analyze spatial patterns of criminogenic events (Bottoms, 2007; Chainey & Tompson, 2008; Santos, 2016; Newton & Felson, 2015; Weisburd, 2015; Roy & Chowdhury, 2023a, b). As might be expected, crime is not evenly distributed across space (Shaw & McKay, 1942; Sherman et al., 1989; Sampson, 2012; Bottoms, 2007; Brantingham & Brantingham, 1981; Brunton-Smith et al., 2013; Roy & Chowdhury, 2023a, b; Biswas & Chatterjee, 2024); rather, it follows a distinct spatial pattern. The crime figures aggregated within a broad spatial context having relevant regional differences (GOSCHIN, 2019) since the major determinants of crime exert differential local effects based on space-specific characteristics. An area's distinct socioeconomic and physical environmental facets might posit the situational backdrop to emerge that area as a hotbed of crime (Santos, 2016). Therefore, focus is needed to identify the neighborhood-level predictors and their variation across space to enable law enforcement officials and policymakers to better understand the geography of crime. Hence allowing targeted policy intervention and allocation of services and resources to curb the issue efficiently. Crime analyses across different countries highlighted various socioeconomic and demographic attributes of crime mechanisms that exacerbate spatial vulnerabilities in connection to the specific socioeconomic and demographic context. These attributes incorporate large-scale resource depletion, poverty, unemployment, socioeconomic deprivation, demographic characteristics, population mobility, weak legal interventions and so on (Sherman et al., 1989; Blau & Blau, 1982; Li & Rainwater, 2000; Baller et al., 2001; Harries, 2006; Messner et al., 2013; Goschin, 2016; Flores & Rodriguez-Oreggia, 2014). Yet, substantially limited studies have focused on assessing the spatial variability of crime and predictors at the local or regional scale. To date, in some criminological studies, the spatial nature of determinants inherent to criminality has largely been accounted (Messner et al., 2013; Sparks, 2011; Ratcliffe, 2010; Beyer et al., 2015; Ingram & Marchesini da Costa, 2017; Gracia et al., 2014). The spatial differentiation of crimes prompts practitioners and researchers from different disciplines to delve into the underlying mechanisms of crimes and discern crime patterns for space specific policy implementation (Anselin et al. , 2000; Andresen & Malleson, 2013; Brantingham & Brantingham, 1984; Biswas & Chatterjee, 2021; Roy & Chowdhury, 2023a, b; Biswas & Das Chatterjee, 2024). Specifically, human geographers prefer to apply cartographic tools to explore spatial context of criminogenic phenomena in a systematic way. (Kabiraj, 2023) whereas, criminologists and sociologists focus on ecological perspectives to elucidate crime phenomena encompassing regional context alongside interpersonal dynamics (Cohen, 1941). Some criminological theories viz., social disorganization theory (Shaw & McKay,1942), routine activity theory (Cohen and Felson, 1979) gained significant importance within the purview of assessing spatiality of crime while considering neighborhood's coherence, social ties, daily routine activities of individual within specific locals (Louderback & Sen Roy, 2018; Andresen, 2006). Due to "diffusion and contagion effects," the spatial interaction of neighborhoods at the local level might worsen the socioeconomic environment of a region and make the region more vulnerable to crime (Flores & Rodriguez-Oreggia, 2014). So, a complete understanding of the changing nature of crime, along with the geospatial factors, is pivotal for implementing space-specific measures to curb criminality at the grassroots level.

Understanding Geography of Crime Against Women in West Bengal

In recent days, the geospatial understanding of the criminogenic issue of crime against women has become the central focus of a substantial number of studies. Most of the studies seeking to unveil the spatial context and shedding light on the underlying mechanisms of crime. In West Bengal, the previous chapter discusses the changing nature of the geography of crime against women. It is obvious from this analysis that crime is not randomly distributed across neighborhoods; rather, specific spatial clustering patterns of different crime categories can be observed from the dynamic understanding of crime against women across space and time. Most of the existing literature emphasizes unraveling the underlying causes of crimes against women in West Bengal. Economic disparities emerge as significant determinants perpetuating discrimination against women as scrutinized in many studies (Ghosh, 2014; Biswas & Chatterjee, 2021; Biswas & Das Chatterjee, 2024). The limited access to education and employment makes women economically dependent and, thus, susceptible to crime. Some studies considered the rate of rapid urbanization and illegal immigration within the state, resulting in high level economic pressure, unemployment, and resource scarcity which fosters social disorganization and increases the risk of women in society (Banerjee, 2011; Karmakar 2013; Chakraborty, 2015; Biswas & Chatterjee, 2024). Socioeconomic crises such as impoverishment, limited availability of basic amenities, and poor accessibility lead to disorganized areas that nurture illicit activities and heighten insecurities among women (Molinari, 2017; Sánchez-Triana et al., 2014; Jana et al., 2013). Deeply rooted patriarchal norms also play a pivotal role in perpetuating gendered discrimination and bringing misfortunes to women (Mukherjee, 2020; Chatterjee, 2021; Biswas & Das Chatterjee, 2023). Moreover, a lack of knowledge about legal aids, fear of the police, ignoring attitude of families, and attached social stigmas fostering the magnitude of women susceptibilities at large (Ghosh, 2014; Chatterjee, 2021; Biswas & Das Chatterjee, 2024). Only a few space-specific case studies on the different forms of crimes against women have been conducted in West Bengal. Biswas et al. (2021) apply geospatial techniques to spatially analyze and map the potential hotspots for eve teasing in a selected urban area. This study identifies some space-specific risk factors and spatially zoned risk-prone areas by integrating GIS with spatial statistics, providing valuable insights for the implementation of necessary measures aimed at securing women's safety in public space. Molinari (2017) scrutinizes how adverse physiographic settings and socioeconomic crises in the Sundarbans region augment high-level social disorganization, cherish organized criminal rackets of trafficking, and exacerbate insecurities among poor rural women. Biswas and Chatterjee (2021) unveil how enormous environmental challenges in the Indian Sundarbans render socioeconomic fragilities and create room for organized criminal activities of trafficking in the remote deltaic villages and increase the susceptibilities of poor women and minors. Prasad (2018) visualizes and analyzes the spatiotemporal evolution of patterns of crimes against women in West Bengal and simultaneously explores their correlations with socioeconomic determinants for a complete understanding of the underlying mechanisms of crime. There have been some other space-specific studies pertaining to crime against women in West Bengal and India that seek to unveil spatial patterns while shedding light on the underlying causation (Niaz, 2003; Jha, 2015; Natarajan, 2016; Mary Santhi & Balaselvakumar, 2018; Vicente et al., 2020; Wani et al., 2020; Chatterjee et al., 2022; Biswas & Das Chatterjee, 2023; Kabiraj, 2023). These studies applied various analytical tools and techniques to scrutinize the geospatial analysis and underlying causation of crimes against women in specific localities. These analyses provide valuable insights into the spatial dynamism of crimes against women and identify geographic areas that require targeted interventions. In addition, understanding the underlying determinants identified in these studies is crucial for implementing effective measures in specific localities. Though there is limited research examining specific aspects of crimes against women, an in-depth understanding is still needed to delve into the spatial heterogeneity of crime against women in West Bengal. As the criminality factors exert differential local effects, giving birth to spatial heterogeneity of crime incidences, therefore, the spatial econometric tools are needed to adopt to estimate the local coefficients specific to each region. Hence, this chapter is going to apply spatial econometric statistics including spatial weights to estimate individual coefficient estimations of determinants of crime against women for each district of West Bengal and predict the future crime rates considering the variation in the relationship among determinants over space. Geographically weighted regression (GWR), considered a valuable instrument to explore such variation over space, is used in this present study. The use of spatial weights in GWR yielding to measure spatial correlations has a strong policy implication with the possibility of far more targeted intervention at the regional level.

Core Risk Factors of Crimes Against Women in West Bengal

Prior to understanding the neighborhood-level determinants of crime against women and their spatial variation, it is necessary to determine the underlying factors responsible for the rising insecurities among women in the crime-ridden districts of West Bengal. For this purpose, principal component analysis (PCA) (Hotelling, 1933) has been performed on available district-wise socioeconomic attributes using SPSS-25 version software to extract the most significant determinants. The socioeconomic attributes considered for performing PCA include population density, sex ratio, decadal growth rate, literacy rate, number of immigrants, nonworking class population (data source: Census of India, 2011), HDI (Human Development Index), GDP (gross domestic product) (data source: Department of Statistics and Programme Implementation, West Bengal 2014–2015), HPI (Human Poverty Index) (data source: District Statistical Handbook), school enrollment ratio (District Statistical Handbook, 2011), and amenities (data source: Economic Review 2017–2018, published by the Department of Planning, Statistics and Programme Monitoring, West Bengal). These socioeconomic attributes have been considered after thoroughly reviewing existing literature on the determinants of crime against women in society.

Data Normality

Prior to performing statistical modeling, data normality was checked to determine whether the available data on socioeconomic attributes satisfy the underlying distribution assumptions. The most frequently made distributional assumption in multivariate statistical techniques is the multivariate normal distribution. A noteworthy property of multivariate normal distribution is that if $X = (X1, X2, X3.......Xn)$ follows a multivariate normal distribution pattern, then all the individual components of $X1, X2, X3.... Xn$ is assumed to be normally distributed. Hence, the data normality of individual Xi needs to be checked to ensure that $X = (X1, X2, X3.......Xn)$ is multivariate normally distributed. In this present study, a QQ (quantile-quantile) plot was used to assess data normality, demonstrating that all the selected attributes are almost normally distributed.

Result of Principal Component Analysis

To determine the number of principal components (PCs), Kaiser's criterion (Dikko et al., 2013) has been followed. The resulting KMO (Kaiser-Meyer-Olkin measure of sampling adequacy) value of 0.76 indicates that the distribution is satisfactory for performing factor analysis. The Bartlett's test of sphericity results (taking 95% confidence level, $\alpha = 0.05$) with a p-value (sig.) of 0.000 is <0.05 and the approximate Chi-square (χ^2) of 96.985 with 45 degrees of freedom (df) strongly reject the null hypothesis (H0) (i.e., $r = 0$) and accept the alternative hypothesis ($H1$) that specifies the existence of noteworthy correlation in-between the variables (Table 4.1). A significance value of Bartlett's test of sphericity (p-value <0.05) indicates that the data do not produce an identity matrix and are satisfactorily multivariate normal and, thus, ideal for performing factor analysis (George and Mallery, 2003). The factor loadings of variables after varimax rotation (Table 4.2) with Kaiser normalization resulting from factor analysis reveals that all 11 attributes are grouped into four major PCs with eigenvalues of more than 1. The PCs are named as (a) socioeconomic adversities and emerging susceptibilities, (b) population influx, gender disparity and social disorganization; (c) sociodemographic indicators and emerging crises, and (d) absence of basic amenities and resulting social disorder respectively after considering the relative proximity of the correlated variables (considered

Table 4.1 KMO and Bartlett's test result

Kaiser-Meyer-Olkin measure of sampling adequacy		0.760
Bartlett's test of sphericity	Approx. Chi-square	96.985
	Df	45
	Sig.	**0.000**

Source: Author's computation

Table 4.2 Factor loading of determinants of crime against women in West Bengal

Variables	Component (rotated component matrix[a])			
	Factor 1	Factor 2	Factor 3	Factor 4
	Socioeconomic adversities and emerging susceptibilities	Population influx, gender disparity, and social disorganization	Sociodemographic indicators and emerging crises	Absence of basic amenities and resulting social disorder
Literacy rate	0.972			
HDI	0.878			
HPI	−0.875			
School enrollment ratio		0.867		
Gender gap		0.610		
Per capita income		−0.592		
No. of immigrants		0.555		
Sex ratio			0.925	
Population density			−0.859	
Availability of basic amenities				−0.911
Eigenvalue	3.148	1.843	1.84	1.256
% of variance	31.478	18.428	18.399	12.558
Cumulative %	31.478	49.906	68.306	**80.863**

Extraction method: Principal component analysis
Rotation method: Varimax with Kaiser normalization

Source: Author's computation
[a]Rotation converged in 6 iterations

significant factor loading of 0.50 and above (Pal & Bagai, 1987). These four factors collectively explained 80.863% of the total variance.

Measures of Collinearity

To estimate the existence of multicollinearity among the independent variables, VIF (variance inflation factor) and tolerance have been checked. Variables having a tolerance value of less than 0.01 and a VIF value greater than 10 denote the existence of strong multicollinearity among the variables (Liu et al., 2003). Table 4.3 reflects

Table 4.3 Collinearity statistics

Predictors	Collinearity statistics	
	Tolerance	VIF (variance inflation factor)
Socioeconomic adversities and emerging susceptibilities	0.601	1.670
Population influx, gender disparity, and social disorganization	0.944	1.020
Sociodemographic indicators and emerging crises	0.939	1.002
Absence of basic amenities and resulting social disorder	0.533	1.710

Source: Author's computation

the collinearity statistics, which reveals that both the tolerance (>0.01) and VIF (<10) values are within satisfactory ranges, suggesting that no multicollinearity exists between the PCs.

Spatial Association of Core Risk Indicators: Results of Geographically Weighted Regression (GWR)

After determining the core risk factors responsible for increasing insecurities among women in West Bengal, it is necessary to understand the spatial association of these determinants since major crime determinants have different local effects. Given the existing spatial heterogeneity, it is unrealistic to expect these factors to have consistent effects across local levels. Understanding how these relationships vary across space has clear policy implications. The application of GWR allows for the assessment of regional variations in the explanatory variables. It accounts for estimating spatial variations in the strength of coefficients by producing estimates of the variables at each data location, factoring in spatial diversity and improving the explanatory power of the model. Details of the GWR model are provided in Chap. 1 of this book.

Results of GWR

In this study, a GWR model was applied using a Gaussian model with an adaptive bi-square geographic kernel to estimate the spatial variations of determinants of crime against women in West Bengal. Standardization of independent variables was performed, and the golden section search method was applied for the selection of optimal bandwidth. The goodness of fit of the model was measured with an Akaike information criterion (AIC) value of 317.793, revealing that the model is a good fit (the smaller the AIC value, the better the fit of the model (Harris, 2016)). The R-squared value of 0.79 also indicates that the model is a good fit and explains around 79% of the total variance in crime occurrence, as shown by this weighted regression equation (Table 4.4).

The coefficient estimates for the GWR model (Table 4.4) demonstrated the best fit. All the independent variables (PCs) proved to be statistically significant in OLS (ordinary least squares) global estimation and remained in the final specification of the GWR model. The Moran's I score (-0.0003, p-value >0.05) in the GWR model residuals signifying that the pattern does not appear to be significantly different from random. This means the GWR model residuals do not suffer from significant spatial autocorrelation. The model outputs of GWR had a condition number less than 30, specifying that the results are reliable (without the existence of strong collinearity). The overall local specificities evolve from the standard residuals of the GWR model (Fig. 4.1), which chooses the standard deviation i.e., the differences between the observed and predicted values of crime in each district specify the existence of strong positive spatial correlation (R-squared 78.98) between all the independent variables and crime rate but with significant variation in the coefficient values. The red colored areas are over predicted areas as showing standard deviation greater than 2.5 means there is a quite large difference between observed and predicted values of crime against women. The local coefficient estimation of each predictor variable is presented in Appendix. The maps evolved from the local coefficient estimations help to capture local specificities of individual factors across districts of West Bengal. Overall, the variable "population influx, gender disparity and social disorganization" has emerged in the model as the most influential predictor of crime against women in West Bengal. The local estimation of this variable ranges from a minimum value of 1025.86 in Darjeeling to a maximum value of 1027.058 in the Purba Medinipur district (Fig. 4.2b). The influence of huge illegal immigration, population influx and resultant social disorganization, high level of gender imparity nurtured in society is maximum in the districts of Purba Medinipur, North-24 Parganas, Paschim Medinipur, Howrah, Hooghly, Bankura, and North-24 Parganas with the highest coefficient values indicating highest crime-enhancing effects of this predictor variable. The variable "socio-economic adversities and emerging susceptibilities" is the second-most influential predictor derived from this model. The local coefficient estimation for this predictor variable ranges from 528.40 in Darjeeling to 528.69 in the South-24 Parganas district (Fig. 4.2a). The influence of socioeconomic adversities is maximum in the districts of South- and North-24

Table 4.4 Results of the GWR model

Variables	Maximum	Minimum
Socioeconomic adversities and emerging susceptibilities	528.692551	528.397964
Population influx, gender disparity, and social disorganization	1027.058086	1025.856639
Sociodemographic indicators and emerging crises	166.361574	165.508415
Absence of basic amenities and resulting social disorder	454.466806	453.966558

Source: Author's computation
Note: R-squared = 0.7897; adjusted R-squared = 0.7295; Moran's I = -0.0003, (p-value = 0.89, >0.05); Akaike AIC = 317.793; sigma =732.5284

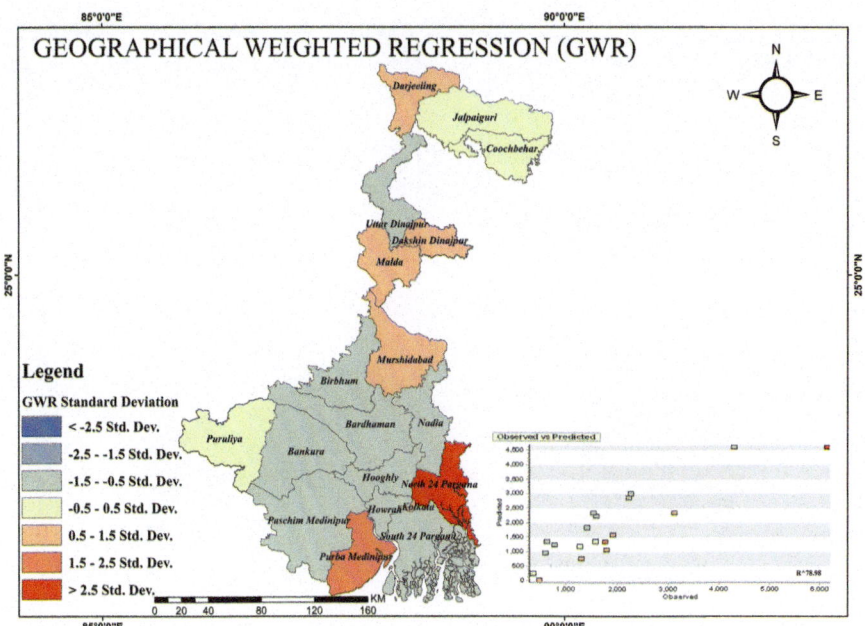

Fig. 4.1 The standard residuals of geographically weighted regression

Parganas, Howrah, Kolkata, Purba Medinipur, Nadia, Hooghly, and Paschim Medinipur, with the highest coefficient values reflecting that this predictor variable has the highest crime-enhancing effects mostly in the southern part of West Bengal. And the intensity of influence gradually fades away towards North Bengal. The local coefficient estimation for the predictor variable "absence of basic amenities and resulting social disorder" ranges from a minimum value of 453.97 in Paschim Medinipur to a maximum of 454.47 in the Jalpaiguri district (Fig. 4.2d). It seems to be the third-most significant predictor variable of crime against women in West Bengal resulting from the GWR model. The influence of this predictor variable is maximum with the highest coefficient values in North Bengal specifically in the districts of Darjeeling, Jalpaiguri, Cooch Behar, Uttar Dinajpur, Dakshin Dinajpur and Malda. The variable "socio-demographic indicators and emerging crises" emerged as a relatively weak predictor as compared to other variables (factors) for increasing insecurities among women in West Bengal (Fig. 4.2c). The local coefficient estimation for this predictor variable ranges from 165.51 in Purba Medinipur to 166.36 in the Darjeeling district. The influence of this predictor variable is also maximum in the districts of Darjeeling, Cooch Behar, Jalpaiguri, Uttar Dinajpur and Dakshin Dinajpur (with the highest coefficient values), indicating that this predictor variable exerts maximum criminality effects in North Bengal and the intensity of influence gradually weakens toward South Bengal.

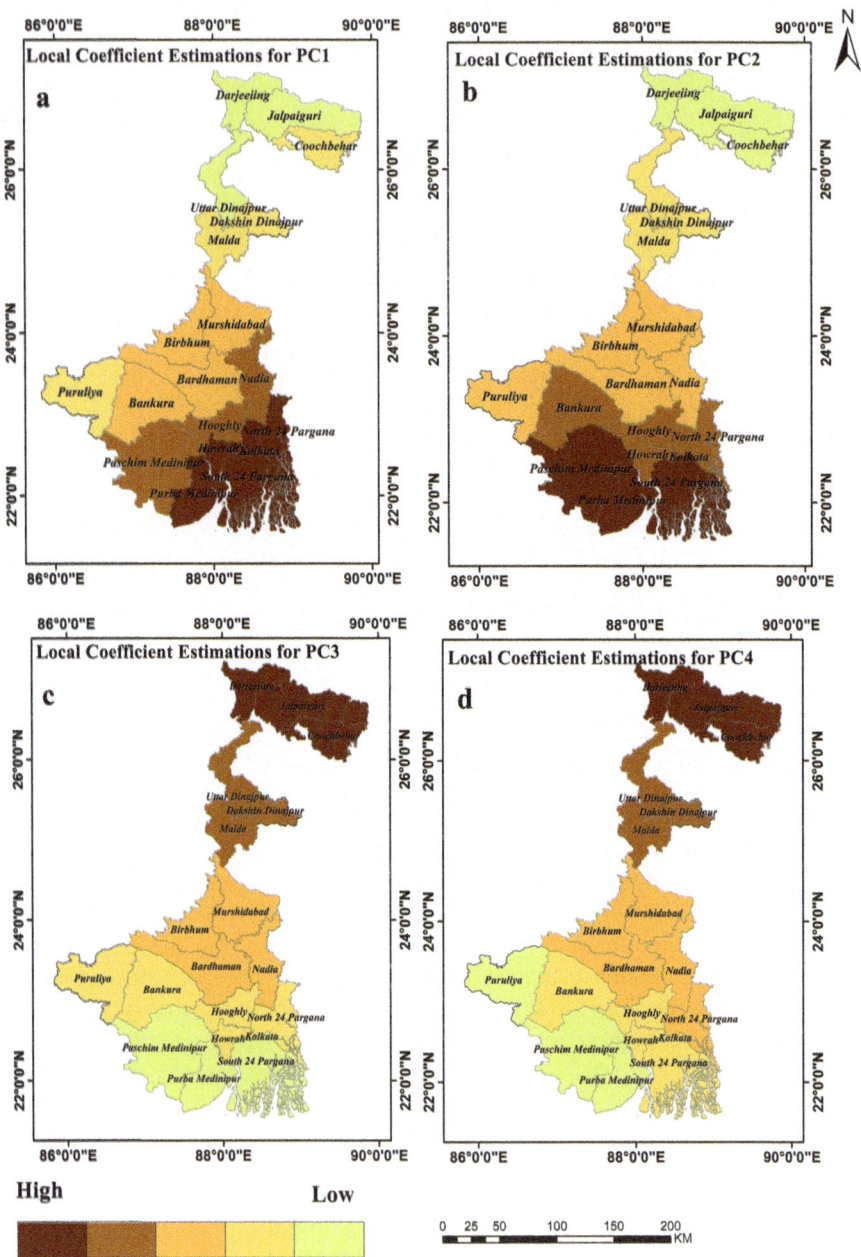

Fig. 4.2 Local coefficient estimations for factors of crime against women in the GWR model: (**a**) PC1, (**b**) PC2, (**c**) PC3, (**d**) PC4

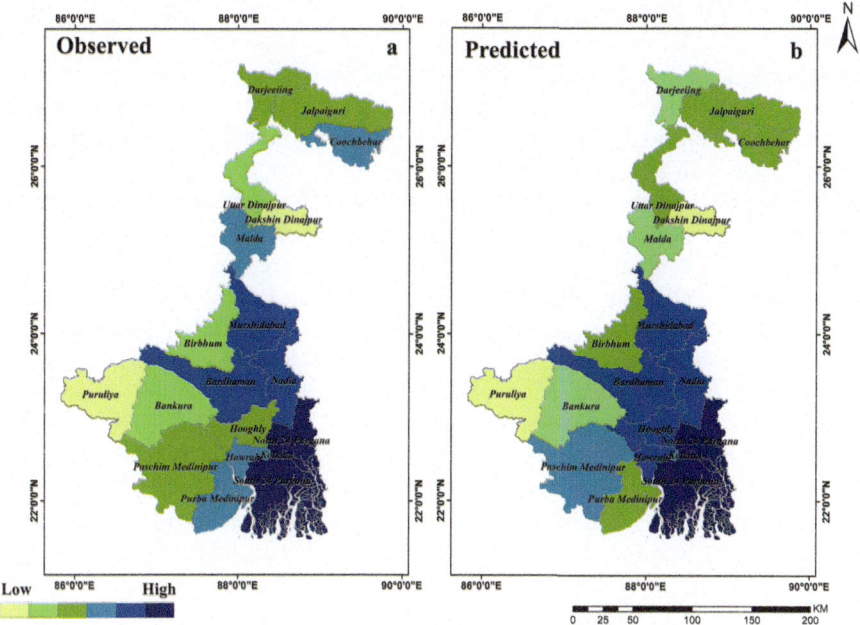

Fig. 4.3 Future potential vulnerable areas of crime against women in West Bengal

Predicting Future Potential Areas of Crime Against Women in West Bengal

As the GWR runs separate regression equations for each district, distinct standardized β coefficients are generated. After transforming the "best-fit" standardized geographically weighted regression into a standardized linear regression equation for each spatial unit, an apparent predictive model of potentially vulnerable districts concerning crimes against women has been obtained (Fig. 4.3). Figure 4.3 shows that the predictor variables can significantly forecast future potential areas, indicating that in the future, the districts of North- and South-24 Parganas, Bardhaman, Nadia, Kolkata, Howrah, Hooghly, Murshidabad, and Paschim Medinipur of South Bengal and Jalpaiguri, Cooch Behar, and Uttar Dinajpur of North Bengal are expected to be more vulnerable to increasing insecurities among women in West Bengal.

Discussions and Conclusion

The present chapter focuses on determining the core risk indicators of crime against women in West Bengal and on applying spatial econometric statistics to estimate how criminality factors exert different local effects, giving birth to regional heterogeneity of crime incidences.

As local anticriminality measures face enormous challenges with regard to accounting for local influencing factors that cross regional boundaries, spatial econometrics offer a proper investigation tool for examining this phenomenon. Hence, this study applies the criminality GWR model to estimate the local determinants that can elucidate the spatial heterogeneity of criminal activities perpetrated against women in West Bengal.

The spatial weights considered in the GWR help make individual coefficient estimations of determinants at the regional level, and based on variations in the relationship among determinants over space, future crime predictions have been made. The results revealed that crime against women in West Bengal is linked to several determinants, which are largely acknowledged in the literature, asserting that high population influx, acute poverty, unemployment, illiteracy, lack of requisite amenities, gender disparity, and other socioeconomic adverse situations account for 80.86% of the variance of women's susceptibilities in West Bengal. Yet significant spatial variabilities of the identified determinants across districts have been observed from local coefficient estimations, which indicates that PC1, PC2, and PC4 are important predictors of crime against women in West Bengal but significantly vary across districts based on local specificity. An apparent predictive model of potentially vulnerable areas of crime in West Bengal has also emerged from the spatial econometric statistics, specifying that all the predictor variables significantly predict that, shortly, the districts of North- and South-24 Parganas, Kolkata, Howrah, Nadia, Bardhaman, Murshidabad, Hooghly, Paschim Medinipur, Uttar Dinajpur, Cooch Behar, and Jalpaiguri might hold foremost positions when it comes to women's susceptibility in West Bengal. Existing literature also demonstrates that socioeconomic adversities, induced by acute poverty, unemployment, and population mobilities, have a strong impact on the increase of criminality in a region (Messner et al., 2013; Goschin, 2016; Debnath & Roy, 2013). The sharing of porous borders with neighboring countries facilitates illegal immigration, creating a population influx situation that exerts enormous economic pressure on the borderland areas of West Bengal, making societies highly disorganized and eventually giving rise to criminal activities in the border areas, putting women and minors at high risk. The cosmopolitan nature of West Bengal with its diversified socioeconomic characteristics and the emergence of a complex social milieu leading to rising insecurities among women in society is also supported by existing studies (Datta, 2005; Kumar, 2016; Chatterjee, 2021; Biswas & Chatterjee, 2021). So, the concerned government should prioritize eliminating poverty, generating employment opportunities to abolish economic deprivation from society and to minimize social disorganization, and restraining the advancement of a criminogenic environment in society. Emphasis need be given on strengthening existing legal systems and causing awareness among people about legal aid. The existing education systems need to be made more value inferring to increase awareness among people about the offenses that women face in their daily lives and make women self-reliant when it comes to their rights, which in turn increases the opportunity cost for criminals. It is very essential to take necessary measures to diminish all traditional patriarchal beliefs of women subjugation and minimize gender disparity in society so that women can live with all majesty and honor. Necessary measures and societal awareness can ameliorate women's positions and dignity in Bengal, paving the way for a crime-free society. This comparative study contributes to an in-depth understanding of the immense spatial differentiation of criminality rates across regional levels, thus enabling law enforcement officials and policymakers to better understand the geography of criminality against women in West Bengal and undertake space-specific policy interventions at the regional level, considering the local influences. Future studies should aim for a more detailed understanding of crime predictors by incorporating both physical and socioeconomic variables and to analyze the spatial variability predictors concerning specific forms of crime against women in West Bengal. They should provide a better understanding of crime hotspots and the local influential factors so that law enforcement agencies can take measures immediately.

Appendix: Local Coefficient Estimations of Variables Based on the GWR Model

Districts	PC1 Socioeconomic adversities and emerging susceptibilities	PC2 Population influx, gender disparity, and social disorganization	PC3 Sociodemographic indicators and emerging crises	PC4 Absence of basic amenities and resulting social disorder
Bankura	528.562468	1026.783666	165.665694	454.020663
Bardhaman	528.575273	1026.713501	165.743811	454.090843
Birbhum	528.539302	1026.583831	165.827851	454.129525
Dakshin Dinajpur	528.499481	1026.238005	166.109282	454.318747
Darjeeling	528.397964	1025.856639	166.361574	454.439288
Purba Medinipur	528.658715	1027.058086	165.508415	453.969517
Howrah	528.64292	1026.935827	165.605848	454.033603
Hooghly	528.622227	1026.847675	165.66674	454.065656
Jalpaiguri	528.445239	1025.918999	166.351614	454.466806
Cooch Behar	528.479767	1026.010188	166.301699	454.453717
Kolkata	528.654818	1026.929151	165.624068	454.056553
Malda	528.489186	1026.293017	166.047553	454.263105
Murshidabad	528.553456	1026.532743	165.889714	454.188102
Nadia	528.607691	1026.701609	165.787696	454.14895
North 24 Pargana	528.660675	1026.881135	165.674557	454.10014
Puruliya	528.520587	1026.749351	165.655146	453.981485
South-24 Pargana	528.692551	1027.049037	165.550957	454.027546
Uttar Dinajpur	528.448686	1026.111553	166.175433	454.332391
Paschim Medinipur	528.611822	1026.963099	165.549129	453.966558

Source: Author's computation

References

Andresen, M. A. (2006). A spatial analysis of crime in Vancouver, British Columbia: A synthesis of social disorganization and routine activity theory. *The Canadian Geographer, 50*(4), 487–502.

Andresen, M. A., & Malleson, N. (2013). Spatial heterogeneity in crime analysis. In M. Leitner (Ed.), *Crime modeling and mapping using geospatial technologies. Geotechnologies and the environment* (Vol. 8). Springer. https://doi.org/10.1007/978-94-007-4997-9_1

Anselin, L., Cohen, J., Cook, D., Gorr, W., & Tita, G. (2000). Spatial analyses of crime. *Criminal Justice, 4*(2), 213–262.

Baller, R. D., Anselin, L., Messner, S. F., Deane, G., & Hawkins, D. F. (2001). Structural covariates of US county homicide rates: Incorporating spatial effects. *Criminology, 39*(3), 561–588.

Banerjee, P. (2011). Bengal-Bangladesh borderland: Chronicles from Nadia, Murshidabad and Maida. In P. Banerjee & A. B. R. Chaudhury (Eds.), *Women in Indian borderlands*. SAGE Publications.

Beyer, K., Wallis, A. B., & Hamberger, L. K. (2015). Neighborhood environment and intimate partner violence: A systematic review. *Trauma, Violence, & Abuse, 16*(1), 16–47.

Biswas, P., & Chatterjee, N. D. (2021). Environmental vulnerability and women trafficking: Exploring the Bengal sundarban deltaic region of India. In *Practices in regional science and sustainable regional development: Experiences from the global south* (pp. 279–296).

Biswas, P., & Chatterjee, N. D. (2024). *Advanced GIS and crime analysis: Violence against women in West Bengal, India*. Taylor & Francis.

Biswas, P., & Das Chatterjee, N. (2023). Tears behind the closed doors: A logistic analysis of domestic violence against women in West Bengal, India. *Crime & Delinquency*.

Biswas, P., & Das Chatterjee, N. (2024). Misery of the dark world: Assessing risk of young women trafficking for commercial sexual exploitation in Darjeeling Himalayas using fuzzy TOPSIS. *SN Social Sciences, 4*(4), 82.

Biswas, P., Das, K., & Das Chatterjee, N. (2021). Analysis of eve-teasing potential zones using geospatial technologies and AHP: A study in Midnapore town, West Bengal, India. *GeoJournal, 86*(3), 1043–1072.

Blau, J. R., & Blau, P. M. (1982). The cost of inequality: Metropolitan structure and violent crime. *American Sociological Review*, 114–129.

Bottoms, A. E. (2007). Place, space, crime, and disorder. In *The Oxford handbook of criminology* (Vol. 4, pp. 528–574).

Brantingham, P. J., & Brantingham, P. L. (1981). *Environmental criminology*. Sage.

Brantingham, P. J., & Brantingham, P. L. (1984). *Patterns in crime*. Macmillan.

Brunton-Smith, I., Sutherland, A., & Jackson, J. (2013). The role of neighbourhoods in shaping crime and perceptions of crime. In *Neighbourhood effects or neighbourhood based problems? A policy context* (pp. 67–87). Springer.

Chainey, S., & Tompson, L. (Eds.). (2008). *Crime mapping case studies: Practice and research*. Wiley.

Chakraborty, S. (2015). Crime against women in India: A socio-economic study. *Asian Journal of Research in Social Sciences and Humanities, 5*(7), 132–141.

Chatterjee, N. D. (2021). *GIS application for crime prediction: A spatio-temporal analysis of crime against women in West Bengal, India*. Notion Press.

Chatterjee, S., Choudhury, S., & Bandyopadhyay, A. R. (2022). Recapitulation of rape: A spatial and temporal analysis of rape in West Bengal, India. *Antrocom: Online Journal of Anthropology, 18*(2).

Cohen, J. (1941). The geography of crime. *The Annals of the American Academy of Political and Social Science, 217*(1), 29–37.

Cohen, L. E., & Felson, M. (1979). Social change and crime rate trends: A routine activity approach. *American Sociological Review*, 588–608.

Datta, P. (2005). Nepali female migration and trafficking. *Journal of Social Sciences, 11*(1), 49–56.

References

Debnath, A., & Roy, N. (2013). Linkage between internal migration and crime: Evidence from India. *International Journal of Law, Crime and Justice, 41*(3), 203–212.

Dikko, H., Osi, A. A., & Bello, Y. (2013). Detecting factors impacting crime rates in Nigeria using principal component analysis. *IOSR Journal of Mathematics, 9*(3), 8–17.

Flores, M., & Rodriguez-Oreggia, E. (2014). Spillover effects of homicides across Mexican municipalities: A spatial regime model approach.

George, D., & Mallery, P. (2003). SPSS for Windows step by step: A simple guide and reference. 11.0 update. https://media.pearsoncmg.com/aw/wps-wraps/retired/. *George 4answers pdf, 549*.

Ghosh, B. (2014). Vulnerability, forced migration and trafficking in children and women: A field view from the plantation industry in West Bengal. *Economic and Political Weekly*, 58–65.

Goschin, Z. (2016). Mapping global and economic crime in Romania. Regional trends and patterns‖. *Romanian Journal of Economics, 42*(2).

GOSCHIN, Z. (2019). Local factors explaining the incidence of criminal offences in Romania. A geographically weighted regression model. *Romanian Statistical Review, 2*.

Gracia, E., López-Quílez, A., Marco, M., Lladosa, S., & Lila, M. (2014). Exploring neighborhood influences on small-area variations in intimate partner violence risk: A bayesian random-effects modeling approach. *International Journal of Environmental Research and Public Health, 11*(1), 866–882.

Harries, K. (2006). Property crimes and violence in United States: An analysis of the influence of population density. *International Journal of Criminal Justice Sciences, 1*(2).

Harris, R. (2016). *Quantitative geography: The basics*. Sage.

Hotelling, H. (1933). Analysis of a complex of statistical variables into principal components. *Journal of Educational Psychology, 24*(6), 417.

Ingram, M. C., & Marchesini da Costa, M. (2017). A spatial analysis of homicide across Brazil's municipalities. *Homicide Studies, 21*(2), 87–110.

Jana, R., Mohapatra, S., & Gupta, A. K. (2013). Integrating climate change adaptation and disaster resilience: Issues for Sundarbans. *Disaster & Development, 7*(1&2).

Jha, D. K. (2015). Geography of rape crime in India: A spatial analysis of official data. *International Journal of Research in Social Sciences, 5*(5).

Kabiraj, P. (2023). Crime in India: A spatio-temporal analysis. *GeoJournal, 88*(2), 1283–1304.

Karmakar, S. (2013). Vulnerable condition of women in south 24 paraganas district, West Bengal. *International Journal of Scientific & Engineering Research, 4*(11), 641–651.

Kumar, A. (2016). Bangladesh border vicinity for human trafficking: An Indian viewpoint. In *Perspectives on India's national security challenges* (pp. 32–42).

Li, J., & Rainwater, J. (2000, June). The real picture of land-use density and crime: a gis application. In *Proceedings of the 20nd annual ESRI international user conference*.

Liu, R. X., Kuang, J., Gong, Q., & Hou, X. L. (2003). Principal component regression analysis with SPSS. *Computer Methods and Programs in Biomedicine, 71*(2), 141–147.

Louderback, E. R., & Sen Roy, S. (2018). Integrating social disorganization and routine activity theories and testing the effectiveness of neighbourhood crime watch programs: Case study of Miami-Dade County, 2007–15. *The British Journal of Criminology, 58*(4), 968–992.

Mary Santhi, P., & Balaselvakumar, K. K. (2018). Geostatistics and geoinformatics in the analysis of crime against women' in Tiruchirappalli city, Tamil Nadu, Geo. *Eye, 7*(2), 32–37.

Messner, S. F., Teske, R. H., Jr., Baller, R. D., & Thome, H. (2013). Structural covariates of violent crime rates in Germany: Exploratory spatial analyses of Kreise. *Justice Quarterly, 30*(6), 1015–1041.

Molinari, N. (2017). Intensifying insecurities: The impact of climate change on vulnerability to human trafficking in the Indian Sundarbans. *Anti-Trafficking Rev, 8*.

Mukherjee, E. B. (2020). Spatial variation of different type of crimes in West Bengal. *International Journal of Multidisciplinary Research and Development, 7*(9), 19–30.

Natarajan, M. (2016). Rapid assessment of "eve teasing" (sexual harassment) of young women during the commute to college in India. *Crime Science, 5*(1), 6.

Newton, A., & Felson, M. (2015). Crime patterns in time and space: The dynamics of crime opportunities in urban areas. *Crime Science, 4*, 11.

Niaz, U. (2003). Violence against women in South Asian countries. *Archives of Women's Mental Health, 6*, 173–184.

Pal, Y., & Bagai, O. P. (1987). A common factory bettery reliability approach to determine the number of interpretable factors. In *IX annual conference of the indian society for probability and statistics held at Delhi, University of Delhi, India*.

Prasad, G. (2018). Analysis of women victimization in West Bengal. *International Journal of Research in Social Sciences, 8*(7), 831–846.

Ratcliffe, J. H. (2010). The spatial dependency of crime increase dispersion. *Security Journal, 23*, 18–36.

Roy, S., & Chowdhury, I. R. (2023a). Geography of crime against women in West Bengal, India: Identifying spatio-temporal dynamics and hotspots. *GeoJournal, 88*(6), 5863–5895.

Roy, S., & Chowdhury, I. R. (2023b). Three decades of GIS application in spatial crime analysis: Present global status and emerging trends. *The Professional Geographer, 75*(6), 882–904. https://doi.org/10.1080/00330124.2023.2223250

Sampson, R. J. (2012). *Great American city: Chicago and the enduring neighborhood effect.* University of Chicago Press.

Sánchez-Triana, E., Paul, T., & Leonard, O. (2014). *Building resilience for sustainable development of the Sundarbans*. The International Bank for Reconstruction and Development, The World Bank.

Santos, R. B. (2016). *Crime analysis with crime mapping*. Sage.

Shaw, C. R., & McKay, H. D. (1942). *Juvenile delinquency and urban areas*. University of Chicago Press.

Sherman, L. W., Gartin, P. R., & Buerger, M. E. (1989). Hot spots of predatory crime: Routine activities and the criminology of place. *Criminology, 27*(1), 27–56.

Sparks, C. S. (2011). Violent crime in San Antonio, Texas: An application of spatial epidemiological methods. *Spatial and Spatio-temporal Epidemiology, 2*(4), 301–309.

Vicente, G., Goicoa, T., Fernandez-Rasines, P., & Ugarte, M. D. (2020). Crime against women in India: Unveiling spatial patterns and temporal trends of dowry deaths in the districts of Uttar Pradesh. *Journal of the Royal Statistical Society Series A: Statistics in Society, 183*(2), 655–679.

Wani, M. A., Shah, S. A., Skinder, S., Dar, S. N., Rather, K. A., Wani, S. A., & Malik, T. A. (2020). Mapping crimes against women: Spatio-temporal analysis of braid chopping incidents in Kashmir Valley, India. *GeoJournal, 85*, 551–564.

Weisburd, D. (2015). The law of crime concentration and the criminology of place. *Criminology, 53*(2), 133–157.

Index

A
ARIMA, 34, 48–52

C
Crime against women, 3–7, 10, 14, 25, 33–52, 66–68, 70–77
Crime hotspots, 1–4, 12, 46, 74, 77
Crime-ridden zone, 33

F
Factor analysis (FA), 14, 15, 69
Fact sheet, 3, 26–31

G
Geographical Weighted Regression (GWR), 13–14, 68, 71–78
Geographic Information Systems (GIS), 3, 10, 14, 33, 35, 51, 65, 67, 75
Geospatial analysis, 3, 67
Geospatial modeling, 3
Geostatistical modeling, 3, 16
GIS environment, 4, 16

I
India, 4, 5, 7, 23–31, 33, 42–45, 55, 63, 67, 68
Integrated data-mining techniques, 3
Intelligence-led policing, 3, 16

M
Moran's I, 11, 12, 34, 44, 46–48, 72

S
Situational environment, 5, 23, 26
Space-based policing, 17, 51
Space-time visualization, 3
Spatial econometric statistics, 68, 75, 76

V
Vulnerable areas, 3, 75, 76
Vulnerable districts, 26, 52, 75

W
West Bengal, 3–7, 12–14, 16, 17, 23–31, 33–63, 66–68, 70–73, 75–77
Women's insecurities, 5, 25, 67, 68, 71, 73, 75

The manufacturer's authorised representative in the EU is Springer Nature Customer Service Centre GmbH, Europaplatz 3, 69115 Heidelberg, Germany. If you have any concerns regarding our products, please contact ProductSafety@springernature.com

Printed and bound by CPI Group (UK) Ltd, Croydon, CR0 4YY

26/03/2026

02078940-0010